TI-83 AND TI-84 MANUAL

KARLA NEAL *Louisiana State University*

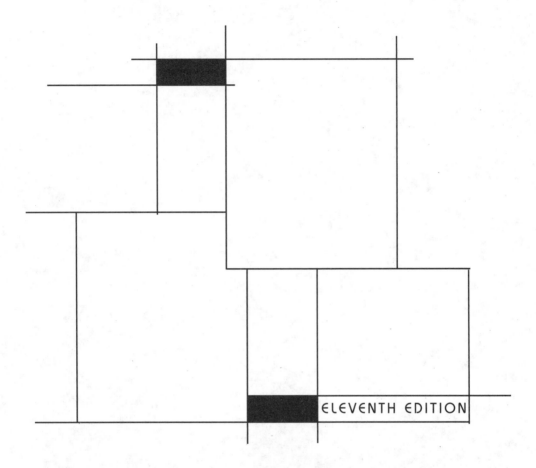

ELEVENTH EDITION

FINITE MATHEMATICS

For Business, Economics, Life Sciences, and Social Sciences

RAYMOND A. BARNETT
MICHAEL R. ZIEGLER
KARL E. BYLEEN

PEARSON

Prentice
Hall

Upper Saddle River, NJ 07458

TECHNOLOGY MANUAL

DALE R. BUSKE *St. Cloud State University*

KARLA NEAL *Louisiana State University*

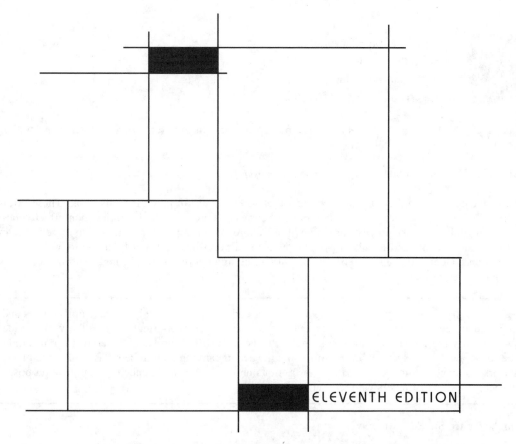

ELEVENTH EDITION

FINITE MATHEMATICS

For Business, Economics, Life Sciences, and Social Sciences

RAYMOND A. BARNETT

MICHAEL R. ZIEGLER

KARL E. BYLEEN

PEARSON

Prentice
Hall

Upper Saddle River, NJ 07458

Vice President and Editorial Director, Mathematics: Christine Hoag
Acquisitions Editor: Chuck Synovec
Supplement Editor: Joanne Wendelken
Project Manager: Robert Merenoff
Senior Managing Editor: Linda Behrens
Supplement Cover Manager: Paul Gourhan
Supplement Cover Designer: Victoria Colotta
Senior Operations Specialist: Ilene Kahn
Senior Operations Supervisor: Diane Peirano

© 2008 Pearson Education, Inc.

Pearson Prentice Hall

Pearson Education, Inc.

Upper Saddle River, NJ 07458

Pearson Prentice Hall™ is a trademark of Pearson Education, Inc.

The author and publisher of this book have used their best efforts in preparing this book. These efforts include the development, research, and testing of the theories and programs to determine their effectiveness. The author and publisher make no warranty of any kind, expressed or implied, with regard to these programs or the documentation contained in this book. The author and publisher shall not be liable in any event for incidental or consequential damages in connection with, or arising out of, the furnishing, performance, or use of these programs.

Printed in the United States of America

10 9 8 7 6 5 4 3 2 1

ISBN 13: 978-0-13-614451-9

ISBN 10:   0-13-614451-9

Pearson Education Ltd., *London*
Pearson Education Australia Pty. Ltd., *Sydney*
Pearson Education Singapore, Pte. Ltd.
Pearson Education North Asia Ltd., *Hong Kong*
Pearson Education Canada, Inc., *Toronto*
Pearson Educación de Mexico, S.A. de C.V.
Pearson Education—Japan, *Tokyo*
Pearson Education Malaysia, Pte. Ltd.

Table of Contents

Chapter 1: Linear Equations and Graphs

For additional help on all aspects of the TI-83 or TI-84 graphing calculator, go to
http://www.prenhall.com/divisions/esm/app/graphing/ti83/

Graphing calculators are particularly efficient tools to use when solving linear equations and inequalities. They can be used to check work as well.

Section 1-1 Linear Equations and Graphs

Example 1: Solving a Linear Equation (page 3)   **Solve and check:** $8x - 3(x-4) = 3(x-4) + 6$

We can use a calculator to check an answer with work that has been done algebraically by storing the value of the variable. In this problem, the solution was found to be $x = -9$. Type in the number -9 in your calculator, press STO, then select the variable X, and press ENTER. After you have done that, you should type in each side of the equation, using the variable X. If the answer is correct, both sides will have the same value.

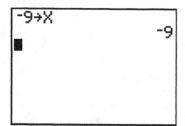
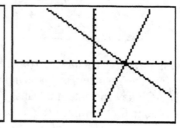

Exercise 1.1.17 (Page 11) Solve $-3(4-x) = 5 - (x+1)$

For this example, we will solve the equation graphically. Enter both equations in the Y= menu. You will have to set the Window so that the point of intersection is shown. It's a good idea to start out using the ZOOM Standard window, and then you can adjust from there. Scroll down to ZStandard and press ENTER. The graphs will be drawn.

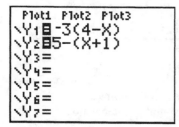

Now you will find the point of intersection, which is the solution to the equation.
Press 2nd TRACE (CALC) and scroll down to 5:intersect, and press ENTER. Move the cursor to near the point of intersection and press ENTER.

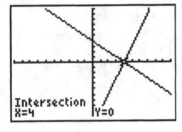
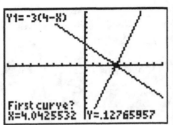
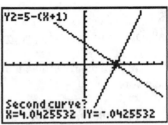

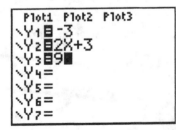 Press ENTER until you see the screen shown.

The solution to the equation is $x = 4$.

Example 6: Solving a Linear Inequality (Page 8) Solve and graph: $2(2x+3) < 6(x-2)+10$

You can check the algebraic work graphically. Enter both sides of the inequality into the Y= menu.
You will have to set the window to show the intersection point. Find the point of intersection. You can see that the point of intersection is 4, and the solution is correct.

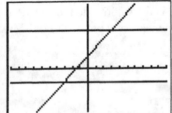

Example 7: Solving a Double Inequality (page 8) Solve and graph $-3 < 2x + 3 \le 9$.

Enter all the parts into the Y= menu, set the window and graph. For this one, we will use TRACE to check the work. Press the TRACE key and enter the value 3 and press ENTER. You will see that the point is the intersection. Do the same for the value -3. You can toggle the cursor to the proper point of intersection.

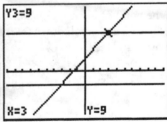

Section 1-2 Graphs and Lines
Example 2 Using a Graphing Calculator (page 16)
Graph $3x - 4y = 12$ and find the intercepts.

Enter $3x/4 - 3$ in the Y= menu. Use the ZOOM 6: Standard window and graph. To find the y-intercept, use TRACE key and enter x=0.

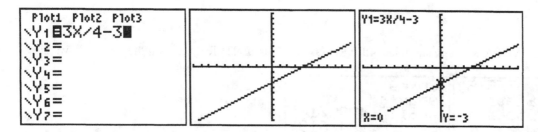

To find the x-intercept, you will use the ZERO function in the CALC menu. Press 2nd TRACE (CALC) and a window will prompt you to enter a left bound. Press ENTER. You will then be prompted to designate a right bound. Move the cursor until it is to the right of the intercept. Press ENTER twice until the ZERO is found.

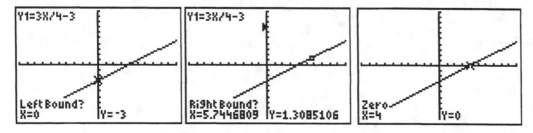

The x-intercept is 4, and the y-intercept is -3.

Section 1-3 Linear Regression

Example 3 Diamond Prices (page 31)
To solve this problem using linear regression requires that the data points be loaded into a list.
Press STAT and then EDIT. Put the numbers in each list. Then enter 2nd Y= to go to STAT PLOT.

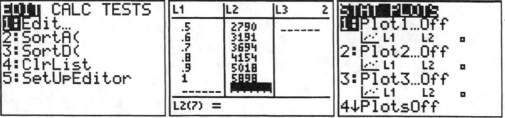

The next window should be configured as shown below on the left. Set the WINDOW to fit the data. Press GRAPH. Make sure no other equations are in the graph menu.

3

 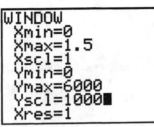 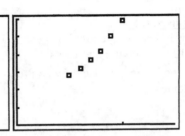

Now go to the STAT CALC menu and select 4:LinReg(ax+b). Press ENTER twice until the screen shows the equation.

```
EDIT CALC TESTS      LinReg
1:1-Var Stats        y=ax+b
2:2-Var Stats        a=6137.428571
3:Med-Med            b=-478.9047619
4:LinReg(ax+b)
5:QuadReg
6:CubicReg
7↓QuartReg
```

Chapter 2: Functions and Graphs

Graphing a function is an important skill that is greatly aided with a graphing calculator.

For additional help on all aspects of the TI-83 or TI-84 graphing calculator, go to
http://www.prenhall.com/divisions/esm/app/graphing/ti83/

Section 2-1 Functions

The TABLE feature is very useful to evaluate the function at a lot of values very quickly. It is more difficult to use it when the relation being graphed is not a function.

Example 1: Point-by-point plotting (page 47). Sketch the graph of (A) $y = 9 - x^2$.

Open the Y= window and type in the function. You will need to set up the TABLE. Press 2nd WINDOW (TBLSET) and make sure your screen looks like the 2nd window shown below. Then go to the TABLE by entering 2nd GRAPH. Enter the values you want to plot. Use the cursor to scroll up and down if you want to change the X values. You can use the values in the TABLE you help you set a window for graphing.

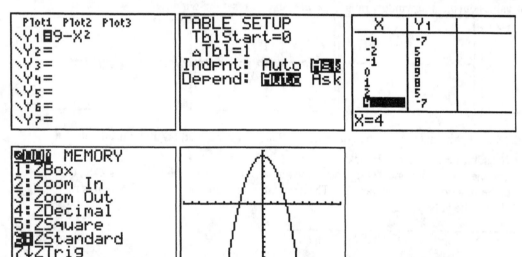

Example 4: Function Evaluation (page 53). If $f(x) = \dfrac{12}{x-2}$ $g(x) = 1 - x^2$ $h(x) = \sqrt{x-1}$ then find

$(A)\ f(6)$ $(B)\ g(-2)$ $(C)\ h(-2)$.

There is more than one way to do this problem on the calculator. What we will do here is show a different method for each part.

$(A)\ f(6)$ For this one, we will use the TABLE function. Enter the function $f(x) = \dfrac{12}{x-2}$ in the Y= menu.

Put parentheses around the denominator. Once the function is entered, access TABLE and type in 6.

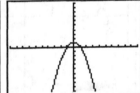 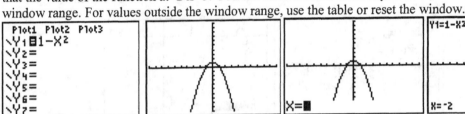 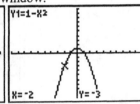

$f(6) = 3$

(B) $g(-2)$ For this problem, we will use the graph to evaluate the function. Enter the function

$g(x) = 1 - x^2$ into the Y= menu, set the window to Standard by pressing ZOOM 6. The graph will be created. Now press 2nd CALC and select the first choice (1:Value). Enter the number -2 and press ENTER. You can see that the value of the function at -2 is -3. You can enter another value if you want, but it must be within the window range. For values outside the window range, use the table or reset the window.

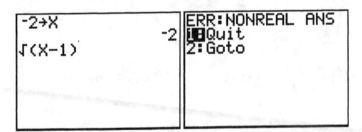

(C) $h(-2)$ We will evaluate this function on the home screen. This is a good method if you need only one value, but is not really useful for multiple values.

On the home screen use the store function (STO) to store the value -2 into the variable X. Press -2 STO X ENTER. Then type in the function using the variable. The calculator will tell you that the function is not defined as a real number at the value x=-2.

```
-2→X
             -2
√(X-1)
```

```
ERR:NONREAL ANS
1∎Quit
2:Goto
```

Section 2-2 Elementary Functions: Graphs and Transformations

A graphing calculator is especially useful in seeing graph transformations.

Example 2: Vertical and Horizontal Shifts (page 66). (A) How are the graphs of the functions
$y = |x|$, $y = |x| + 4$, $y = |x| - 5$ **related? (B) How are the graphs of** $y = |x|$, $y = |x + 4|$, $y = |x - 5|$ **related?**

Enter the three functions in the Y= menu. Set the WINDOW to ZOOM Standard and compare the graphs. The absolute value function is found under MATH NUM window. You should access that from the Y= window. Just press ENTER and it will be transferred to the graph window. You can use the TRACE key to note values.

(A) $y = |x|$, $y = |x| + 4$, $y = |x| - 5$

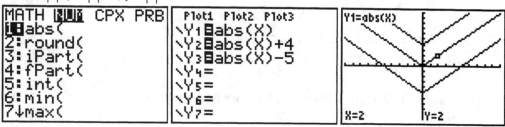

(B) $y = |x|$, $y = |x+4|$, $y = |x-5|$

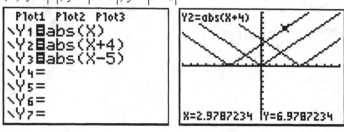

Example 4: Reflections, Stretches, and Shrinks (page 68). (A) How are the graphs of the functions $y = |x|$, $y = 2|x|$, $y = 0.5|x|$ related? (B) How are the graphs of $y = |x|$, $y = -2|x|$ related?

Enter the functions as before. Use the TRACE key to toggle back and forth between the functions so that you can see the function used to create that graph.

(A) $y = |x|$, $y = 2|x|$, $y = 0.5|x|$

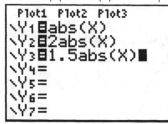
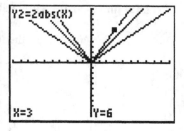

(B) $y = |x|$, $y = -2|x|$

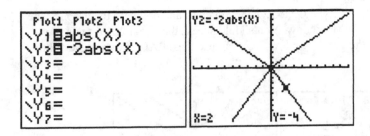

Example 5: Reflections, Stretches, and Shrinks (page 69). How are the graphs of the functions $y = |x|$ **and** $y = -|x-3|+1$ **related?**

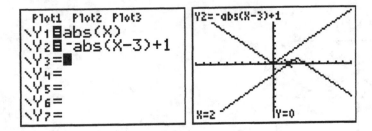

Section 2-3 Quadratic Functions

The example below introduces several ways that a calculator can be used to solve equations and inequalities.

Example 1: Intercepts, Equations, and Inequalities (page 76).
(C) Graph $f(x) = -x^2 + 5x + 3$ **in a standard viewing window.**
(D) Find the x **and** y **intercepts to four decimal places using trace and the zero command.**
(E) Solve the quadratic inequality $-x^2 + 5x + 3 \geq 0$ **graphically to four decimal places.**
(F) Solve the equation $-x^2 + 5x + 3 = 4$ **to four decimal places graphically using intersection.**

(C) Enter the function and use ZOOM:6 as the window. Make sure you use the negative sign and not the subtraction sign.

(D) Find the x and y intercepts to four decimal places using trace and the zero command.
The y-intercept is the value of the function when $x=0$. With the graph screen showing, press TRACE. The value at the center of the screen will be shown. In this case, it is for $x=0$. We see that the value of the y-intercept is 3.

8

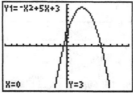

To find the x-intercepts, we will use the ZERO command found in the CALC menu. From the graph screen press 2nd TRACE and select 2:zero. Move the cursor until it is below the first intercept and then press ENTER. Now move the cursor until it is above the intercept and then press ENTER twice.

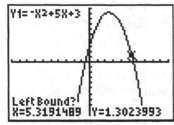

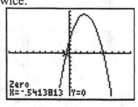

Repeat the process for the other zero.

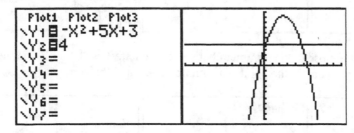

(E) Solve the quadratic inequality $-x^2 + 5x + 3 \geq 0$ graphically to four decimal places.
To solve this inequality, simply use the values found in part (D) to form the interval where the function lies above the x-axis. From this you can determine that $-x^2 + 5x + 3 \geq 0$ over the interval [-0.851, 5.541]

(F) Solve the equation $-x^2 + 5x + 3 = 4$ to four decimal places graphically using intersection.
Enter the function y=4 into the Y= menu and graph.

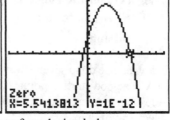

To find the points of intersection, press 2nd TRACE and select 5:intersect. Move the cursor until it is close to one point of intersection. Press ENTER until you see the point of intersection.

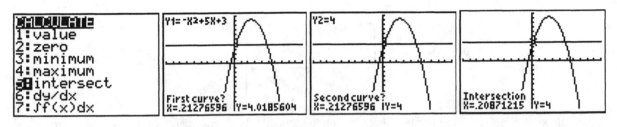

Repeat to find the other point of intersection.

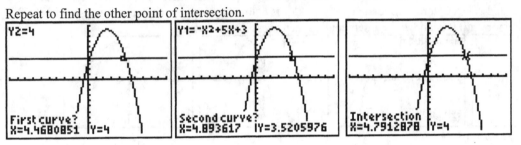

Example 2: Analyzing a Quadratic Function (page 81). Given the quadratic function
$f(x) = 0.5x^2 - 6x + 21$ **(E) Graph the function using a suitable viewing window and (F) Find the vertex and the maximum or minimum.**

(E) Finding the window can be done several ways. The TABLE is useful, or you can use TRACE. The STANDARD window is a good starting point. For this function, this will be sufficient to use for finding the vertex. Once the vertex can be seen, you can find the coordinates of the vertex using the CALC menu. For this function, we will select 3:minimum.

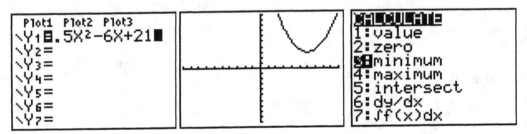

Move the cursor until it is close to the minimum but still to the left. Press ENTER and then move to a right bound. Press ENTER until you see the point. The minimum is 3 and the vertex is (6,3).

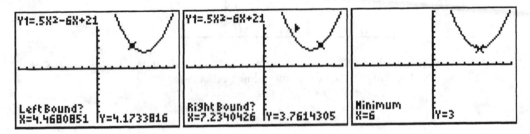

Example 5: Outboard Motors (page 86).

To solve this problem using quadratic regression requires that the data points be loaded into a list.
Press STAT and then EDIT. Put the numbers in each list. Then enter 2nd Y= to go to the STAT PLOT.

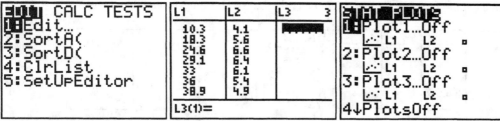

The next window should be configured as shown below on the left. Set the WINDOW to fit the data.
Press GRAPH. Make sure no other equations are in the graph menu.

Now go to the STAT CALC menu and select 5:QuadReg. Press ENTER twice until the screen shows the
equation. Graph the function and use trace to find other points. To have the regression equation automatically
entered into the Y= menu, from the Y= menu, go to the VARS menu, select 5:Statistics, then select EQ and
1:RegEQ and press ENTER.

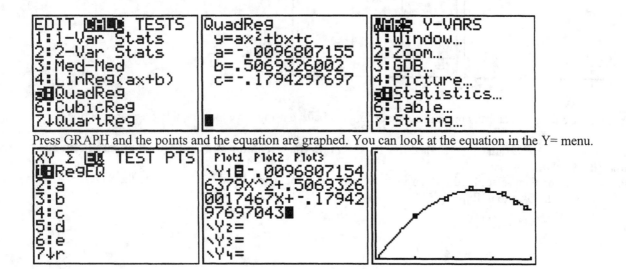

Press GRAPH and the points and the equation are graphed. You can look at the equation in the Y= menu.

Section 2-4 Exponential Functions

Example 1: Graphing Exponential Functions (page 95) Sketch the graph of $y = \left(\frac{1}{2}\right)4^x$, $-2 \le x \le 2$.

Enter the function into the Y= menu and set the window. To determine the values for the window, use the table. You can see the values of the function for the given interval of x. Set the window and then graph.

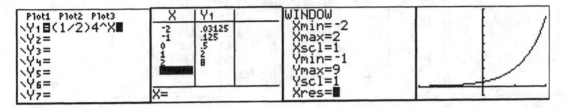

Example 2: Exponential Growth (page 98). For $N = N_0 e^{1.386t}$, **find the number of bacteria present in (A) 0.6 hour and (B) in 3.5 hours if** $N_0 = 25$.

This can be done using the TABLE function. Enter the function into the graph Y= menu. Go to the table and enter the values 0.6 and 3.5. It is not necessary to graph the function to evaluate it using the TABLE, so you do not need to set a window.

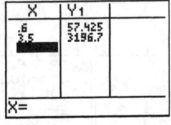

There are 57 bacteria present when x=0.6 and 3197 bacteria present when x=3.5.

You can use the calculator to approximate an exponential curve using the exponential regression function.

Example 4: Depreciation (page 100). Enter the data into the calculator and find the exponential function that fits the data.
Press STAT and then select 1:Edit. Enter the data values into the lists L1 and L2. Now go to the STAT: CALC menu and scroll down to 0: ExpReg

Press ENTER twice until the screen below is seen. You can graph the data points and the exponential regression function. To graph the data points, press 2nd Y= which is the STAT PLOT menu. Press ENTER and access the screen. It should be set up as shown.

Set the window. To have the regression equation automatically entered into the Y= menu, from the Y= menu, go to the VARS menu, select 5:Statistics, then select EQ and 1:RegEQ and press ENTER.

Now press graph and the two graphs are plotted.

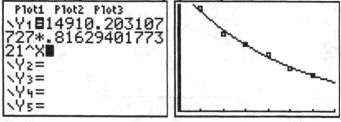

Section 2-5 Logarithmic Functions

Your calculator will evaluate logarithms of base 10 and base e. These are the keys LOG and LN.

Example 7: Calculator Evaluation of Logarithms (page 111). Use a calculator to evaluate each to six decimal places: (A) log 3,184 (B) ln 0.000349 (C) log (-3.24)

These can all be evaluated on the home screen. Make sure you close the parentheses. After you press ENTER for log (-3.24), you will get the error message shown below since that value is not defined.

```
log(3184)
          3.502973059
ln(.000349)
          -7.960438636
log(-3.24)
```

```
ERR:NONREAL ANS
1■Quit
2:Goto
```

Logarithmic equations can be solved graphically.

Example 8: Solving log$_b$x=y for x. (page 112). Find x to four decimals for log x = -2.315.

Enter two functions into the Y= menu. Set a window, graph, and find the point of intersection.

```
Plot1 Plot2 Plot3
\Y1■log(X)
\Y2■-2.315■
\Y3=
\Y4=
\Y5=
\Y6=
\Y7=
```

```
WINDOW
Xmin=0
Xmax=2
Xscl=1
Ymin=-4
Ymax=2
Yscl=1
Xres=1
```

```
Intersection
X=.00484172  Y=-2.315
```

Example 11: Home Ownership Rates (page 115). Use logarithmic regression to find the best model in the form y=a+b ln x.

As with Example 4 on page 100, we enter the data into the list, find the regression equation, enter the regression equation into the Y= menu and graph the data and the equation.

Let x=0 represent the year 1900.

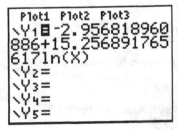

```
L1      L2      L3    2
50      55      ------
60      61.5
70      62.5
80      64.4
90      64.2
100     67.4
------  -----
L2(7) =
```

```
EDIT ■■■■ TESTS
3↑Med-Med
4:LinReg(ax+b)
5:QuadReg
6:CubicReg
7:QuartReg
8:LinReg(a+bx)
9■LnReg
```

```
LnReg
y=a+blnx
a=-2.956818961
b=15.25689177
```

Once you have set the window and graphed the function, use the VALUE function (2nd CALC 1) to predict that home ownership in 2015 will be 69.4%.

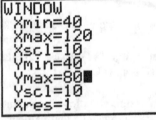

```
Plot1 Plot2 Plot3
\Y1■-2.956818960
886+15.256891765
6171ln(X)
\Y2=
\Y3=
\Y4=
\Y5=
```

```
WINDOW
Xmin=40
Xmax=120
Xscl=10
Ymin=40
Ymax=80■
Yscl=10
Xres=1
```

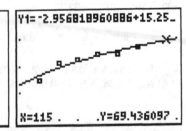

```
Y1=-2.956818960886+15.25_
```
```
X=115 .    .Y=69.436097 .
```

Chapter 3: Mathematics of Finance

For additional help on all aspects of the TI-83 or TI-84 graphing calculator, go to
http://www.prenhall.com/divisions/esm/app/graphing/ti83/

<u>Section 3-1 Simple Interest</u>

Example 2: Present Value of an Investment (page 129). If you want to earn an annual rate of 10% on your investments, how much (to the nearest cent) should you pay for a note that will be worth $5000 in 9 months?

The equation is $A = P(1+rt)$, and we have $5000 = P(1+0.1(0.75)) \Rightarrow P = \dfrac{5000}{(1+0.1(0.75))}$.

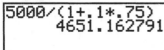

Make certain that you put parentheses around the denominator.

Exercise 21: (page 132) A=$14,560; P=$13,000; t=4 months; r = ?

The formula is $A = P(1+rt)$, and we have $r = \dfrac{A-P}{Pt} = \dfrac{14560-13000}{13000\left(\dfrac{4}{12}\right)}$.

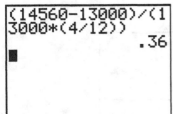

Pay attention to the parentheses around the numerator and the denominator.

<u>Section 3-2 Compound and Continuous Interest</u>

Example 2: Compounding Daily and Continuously (page 138). What amount will an account have after 2 years if $5000 is invested at an annual rate of 8% (A) compounded daily? (B) compounded continuously?

(A) Using the compound interest formula $A = P\left(1+\dfrac{r}{m}\right)^{mt}$, we have $A = 5000\left(1+\dfrac{.08}{365}\right)^{(365 \cdot 2)}$. Be careful to place parentheses in the proper places, especially the exponent.

15

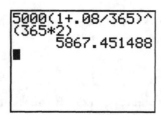

(B) Use the continuous compound interest formula $A = Pe^{rt}$. This is an easier formula to enter into the calculator. Don't forget the parentheses in the exponent. $A = 5000e^{(0.08 \cdot 2)}$.

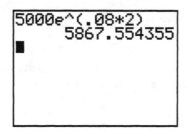

Example 5: Computing Growth Time (page 142). How long will it take $10,000 to grow to $12,000 if it is invested at 9% compounded monthly?

We are solving the equation $12000 = 10000\left(1 + \dfrac{.09}{12}\right)^n$. We will solve this problem graphically. Enter both sides

of the equation into the Y= menu. Set a proper window that will include the point of intersection.
Graph and find the point of intersection. Intersection is found by pressing 2nd TRACE 5. Move the cursor near the point of intersection and press ENTER until you see the screen shown below.
We can see that it will take about 25 months.

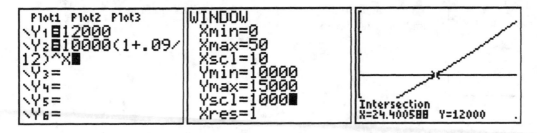

There is another solution method using the Equation Solver. This is found under the MATH menu.
Press ENTER and the Equation Solver will come up. Enter the equation using the ALPHA key.

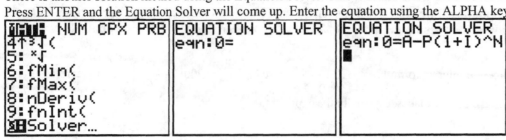

Press ENTER and then put in the values. Leave the value for N blank and then hit ENTER.

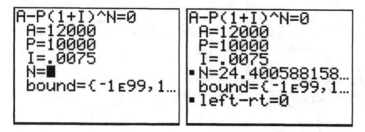

Section 3-3 Future Value of an Annuity; Sinking Funds

Example 4: Approximating and Interest Rate (page 155). Find the interest rate, *i*, if FV=$160,000, PMT=$100, and *n*=360.

This data results in the equation $160,000 = 100\dfrac{(1+i)^{360}-1}{i}$. This equation can be solved using graphs and the

intersect function. It can also be solved using the equation solver which was demonstrated in section 3-2, example 5.

However, there is another method that can be used with the TI-84. It is the TVM Solver.

Press APPS and select 1:Finance. The TVM Solver is the first choice. Enter the values shown, leaving the value for I% blank. Press ALPHA and ENTER to solve for I. You may have to put a 0 in for I% until the other values are entered. Then go back and clear the 0.

Section 3-4 Present Value of an Annuity; Amortization

Example 2: Retirement Planning (page 160).
What we are trying to do here is to find out how much money to deposit annually for 25 years to receive 20 payments of $25,000 if the interest rate is 6.5% compounded annually.

Find PV with PMT= $25,000, i=0.065, and n=20. Access the TVM Solver as demonstrated in Example 4 from section 3-3. Enter the values and then press ALPHA and ENTER. The amount that will have to be in the account is $275462.68

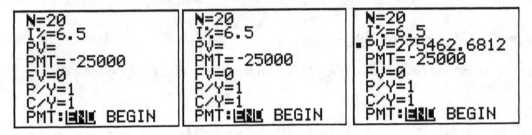

Now that we have the amount that will have to be in the account in 25 years, we will find the amount of 20 annual deposits that will yield that amount. Enter the values as shown and press ALPHA and ENTER. We now have that an annual payment of $4677.76 for 25 years, will provide a $25000 a year for 20 years.

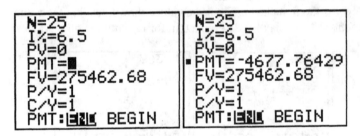

Chapter 4: Systems of Linear Equations; Matrices

For additional help on all aspects of the TI-83 or TI-84 graphing calculator, go to
http://www.prenhall.com/divisions/esm/app/graphing/ti83/

Section 4-1 Review: Systems of Linear Equations in Two Variables

Example 3: Solving a System Using a Graphing Calculator (page 179).

To solve a system by using a graphing calculator, you will have to put both equations into the Y= menu. Then you need to set an appropriate window. If you don't have an idea of the window to use, you can try the ZOOM 6:ZStandard window. Or, you can use the TABLE to locate values of x where the two equations are close in value. Once the window is established, you will then press 2nd TRACE (CALC) 5:Intersect. Move the cursor until it is close to the point of intersection and press ENTER until the intersection point is shown.

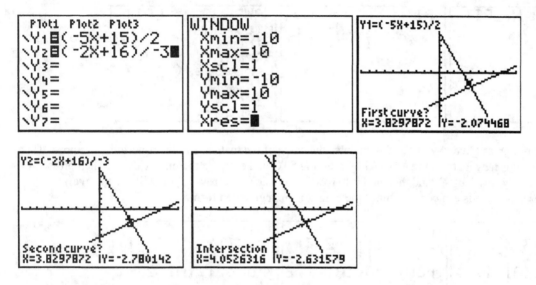

Section 4-2 Systems of Linear Equations and Augmented Matrices

Solving systems of equations can often be solved using augmented matrices. You can command the calculator to perform row operations.

Example 1: Solving a System Using Augmented Matrix Methods (page 191).

To access the Matrix menu, press 2nd and x⁻¹. You will select one of the matrices to enter the values. For this example, we will use matrix [A]. You will have to move the cursor to EDIT to enter the values. The matrix will appear blank unless it has already been edited. Change the dimensions of the matrix to 2 X 3. Move the cursor over the numbers 1 X 1 and change it to 2 X 3. Use the down cursor to move into the matrix and enter the values. As you enter values and press ENTER, the cursor will move to the next space.

NAMES MATH **EDIT**	MATRIX[A] 1 ×1	MATRIX[A] 2 ×3
1:[A]	[0]	[3 4 1]
2: [B]		[1 -2 **7**]
3: [C]		
4: [D]		
5: [E]		
6: [F]		
7↓[G]		2,3=7

To perform row operations, you first return to the home screen and call up the matrix [A]. Access the MATRIX menu again and press 1. Press ENTER and the matrix name will appear on the home screen. Press ENTER again and you will see the matrix [A]. To perform a row operation, access the matrix menu, move to MATH and scroll down to the row operations.

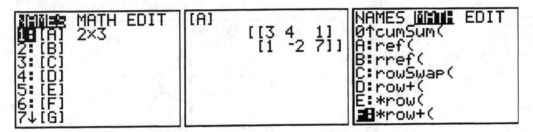

NAMES MATH EDIT	[A]	NAMES **MATH** EDIT
1:[A] 2×3	[[3 4 1]	0↑cumSum(
2: [B]	[1 -2 7]]	A:ref(
3: [C]		B:rref(
4: [D]		C:rowSwap(
5: [E]		D:row+(
6: [F]		E:*row(
7↓[G]		**F:***row+(

To swap two rows, press **C:rowSwap(** and then enter the name of the matrix, and the rows to be swapped. Then you need to store this new matrix in matrix [A] by pressing STO and entering the matrix name. Then press ENTER. To multiply row 1 by -3, and then add it to row 2 to create a new row 2, you will use the ***row+(** command. These commands can continue until the matrix is in augmented form.

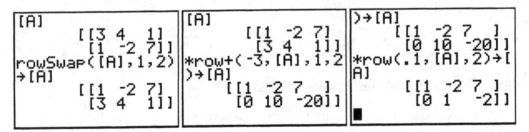

```
[A]
          [[3 4  1]
           [1 -2 7]]
rowSwap([A],1,2)
→[A]
          [[1 -2 7]
           [3 4  1]]
```

```
[A]
          [[1 -2 7]
           [3 4  1]]
*row+(-3,[A],1,2
)→[A]
          [[1 -2 7 ]
           [0 10 -20]]
```

```
)→[A]
          [[1 -2 7 ]
           [0 10 -20]]
*row(.1,[A],2)→[
A]
          [[1 -2 7 ]
           [0 1 -2]]
```

The augmented matrix is now available. We can see the solution $x_1=3$ and $x_2=-2$.

```
[A]
          [[1 -2 7 ]
           [0 1 -2]]
*row+(2,[A],2,1)

          [[1 0 3 ]
           [0 1 -2]]
```

Section 4-3 Gauss-Jordan Elimination

The TI-84 will put a matrix in row echelon form and reduced row echelon form. This makes solving a system of equations very easy.

Example 2: Solving a System Using Gauss-Jordan Elimination (page 200).

Create the matrix for the system. You will notice that all of the data will not be visible for larger matrices, but you will be able to see this one on the home screen. Store it in [A].

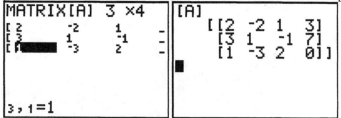

Go into the Matrix menu and select **rref** in order to put the matrix in reduced row echelon form. Notice that one of the entries is a decimal. In order to view this element, scroll to the right. The last number in the second row is 0.

```
NAMES  MATH  EDIT
0↑cumSum(
A:ref(
B:rref(
C:rowSwap(
D:row+(
E:*row(
F:*row+(
```

```
rref([A])→[A]
[[1 0 0 2
 [0 1 0 -3.5E-1…
 [0 0 1 -1
                ...
```

```
[[1 0 0 2       …
 [0 1 0 -3.5E-1…
 [0 0 1 -1      …
[A]▶Frac
    0 0 2        ]
    1 0 -3.5E-13]
    0 1 -1       ]]
```

Example 3: Solving a System Using Gauss-Jordan Elimination (page 201).

How can you tell when a system has to no solution using the calculator? You will go through the same steps as in the last example.

```
[[2  -4 1  -4]
 [4  -8 7  2 ]
 [-2 4  -3 5 ]]
rref([A])→[A]
[[1 -2 0 0]
 [0 0  1 0]
 [0 0  0 1]]
```

The last row is equivalent to:
$$0x_1 + 0x_2 + 0x_3 = 1.$$

This is obviously a false statement, so we know that the system has no solution.

Example 4: Solving a System Using Gauss-Jordan Elimination (page 202).
How can we tell when a system has infinite solutions? This example will demonstrate this. Enter the matrix as before and put it in reduced row echelon form "rref".

```
[[3   6   -9  15]
 [2   4   -6  10]
 [-2 -3  4   -6]]
rref([A])→[A]
   [[1  0  1   -3]
    [0  1  -2  4 ]
    [0  0  0   0 ]]
```

The resulting matrix yields the system
$$x_1 \quad + x_3 = -3$$
$$x_2 - 2x_3 = 4$$
From this the solution to the system can be found.

Section 4-4 Matrices: Basic Operations

The basic operations of addition, subtraction, scalar multiplication, and matrix products can be performed on the calculator.

Example 1: Matrix Addition

Perform the addition: $\begin{bmatrix} 2 & -3 & 0 \\ 1 & 2 & -5 \end{bmatrix} + \begin{bmatrix} 3 & 1 & 2 \\ -3 & 2 & 5 \end{bmatrix}$ by entering each matrix. Then enter the name of each matrix with the addition operation.

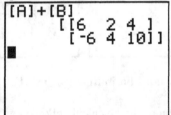

The same procedure is used for subtraction of matrices.

Example 4: Multiplication of a Matrix by a Number (page 213).

Enter the matrix and then multiply by -2.

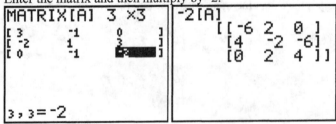

Example 8: Matrix Multiplication (page 216).
To multiply two matrices, they must have the proper dimensions. If they do not, the calculator will show an error message.

Enter the two matrices and then multiply them together.

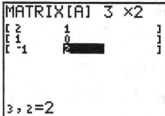 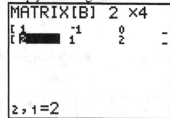 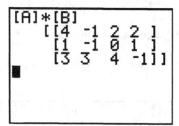

Section 4-5 Inverse of a Square Matrix

If a matrix has an inverse, it can be found on the calculator. The process can be quite long when done by hand, so the calculator can be a real time saver.

Example 2: Finding the Inverse of a Matrix (page 228).
Find the inverse of the matrix.

$$M = \begin{bmatrix} 1 & -1 & 1 \\ 0 & 2 & -1 \\ 2 & 3 & 0 \end{bmatrix}$$

Enter the matrix into the calculator. Call up the matrix on the home screen and then press the x^{-1} key on the calculator. The inverse of the matrix will be displayed.

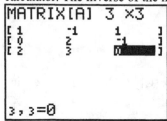

Example 4: Finding a Matrix Inverse (page 232).
If the calculator does not have an inverse, an error message will be displayed.

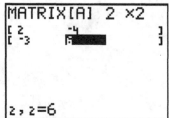 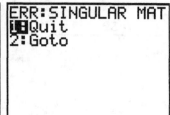

Section 4-6 Matrix Equations and Systems of Linear Equations

The methods shown in this section are primarily for calculations done by hand. The ability of a graphing calculator to put a matrix into reduced form, negates the need to find the inverse of a matrix to solve a system. Some of the homework problems can be done using the graphing calculator.

Exercise 13 (page 242) Find x_1 and x_2.

$\begin{bmatrix} 1 & -1 \\ 1 & -2 \end{bmatrix} \begin{bmatrix} x_1 \\ x_2 \end{bmatrix} = \begin{bmatrix} 5 \\ 7 \end{bmatrix}$ This uses the property $AX = B \rightarrow X = A^{-1}B$. We will use $A = \begin{bmatrix} 1 & -1 \\ 1 & -2 \end{bmatrix}$, $B = \begin{bmatrix} 5 \\ 7 \end{bmatrix}$.

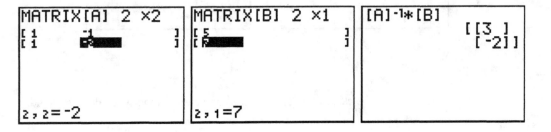

Section 4-7 Leontief Input-Output Analysis

Example 1: Input-Output Analyis (page 249)
The rather involved use of matrices in this problem can be done quickly with the calculator.
You will have to store the matrices. Store the identity matrix, I, in [A]. Store matrix M in [B].

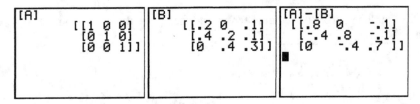

Create matrix [C] by the subtraction [A]-[B] (I-M). Then create the inverse matrix [C]⁻¹ ([I-M]⁻¹) and store it in [D]. Create matrix [E]. The solution is found by multiplying [D][E].

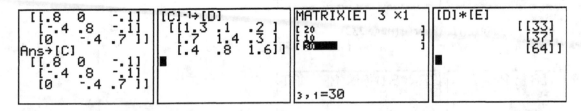

Chapter 5: Linear Inequalities and Linear Programming

For additional help on all aspects of the TI-83 or TI-84 graphing calculator, go to
http://www.prenhall.com/divisions/esm/app/graphing/ti83/ There is a section on using the Inequality Graphing
Application.

Section 5-1 Inequalities in Two Variables

The TI-84 will shade an area above or below a line.

Example 1: Graphing a Linear Inequality (page 261). Graph $2x - 3y \leq 6$.

First you must solve the inequality for y in order to put it into the Y= menu. $\dfrac{2x}{3} - 2 \leq y$.

Enter this into the Y= menu. Set the window to ZOOM 6, and select GRAPH.

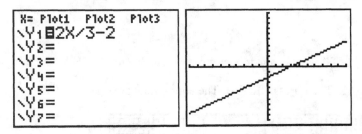

After inserting the test point (0,0), it is determined that the point is in the solution set. In order to shade the area
above the line, go back to the Y= menu and move the cursor to the left of Y_1. The line will start flashing.
Hit the ENTER key until you see the screen below. The triangle indicates that the area above the line will be
shaded. The shaded area is the solution to the inequality.

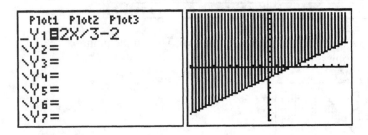

Section 5-2 Systems of Linear Inequalities in Two Variables

The TI-84 has a built in application that will graph and shade the solution area of a system of inequalities.
Select the APPS key and scroll to the application **Inequalz**. Open the application and you will be taken to the
graph editor. Move the cursor over the = sign. The symbols in the menu at the bottom are accessed via F1-F5
by pressing ALPHA and then the number you wish.

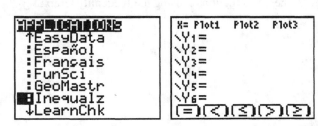
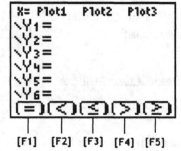

We will now solve a system using this application.

Example 1: Solving a System of Linear Equations Graphically (page 268).

Solve the system $\begin{aligned} x+y &\ge 6 \\ 2x-y &\ge 0 \end{aligned}$.

Solve each for y. $\begin{aligned} y &\ge 6-x \\ y &\le 2x \end{aligned}$. Enter each inequality. Move the cursor over the = sign and enter ALPHA F5 to select ≥ for Y1 and ALPHA F3 to select ≤ for Y2. Then press GRAPH. The appropriate areas are shaded. Press ALPHA F1 to access the shades menu.

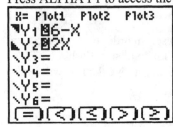
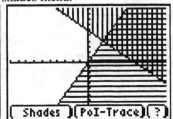
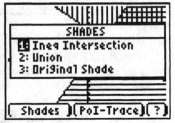

Select the first choice 1:Ineq Intersection to show only the intersection of the two. To find the point of intersection, press ALPHA F3. The intersection point is shown.

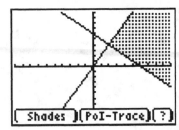
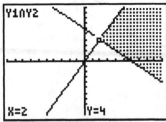

Example 2: Solving a System of Linear Inequalities Graphically (page 269).

We can solve the system of inequalities as in the last example. Take each inequality and simplify. Enter each inequality into the Y= menu, except for $x \geq 0$. To e nter that inequality, move the cursor to the top left of the screen where you see the X=. Press ENTER and then put $x \geq 0$ into that menu.

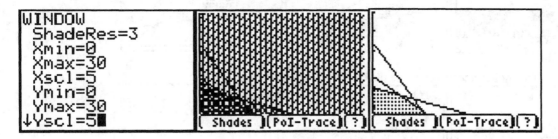

$$\begin{aligned} 2x + y &\leq 22 \\ x + y &\leq 13 \\ 2x + 5y &\leq 50 \\ x &\geq 0 \\ y &\geq 0 \end{aligned} \qquad \rightarrow \qquad \begin{aligned} y &\leq 22 - 2x \\ y &\leq 13 - x \\ y &\leq 10 - 2x/5 \\ x &\geq 0 \\ y &\geq 0 \end{aligned}$$

Set the window and then press GRAPH. Now you will shade the intersection. Press ALPHA F1 and select Intersection.

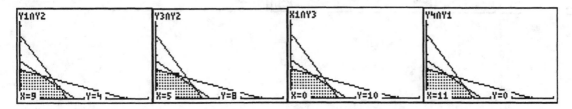

Now select ALPHA F3 and use the up and down cursor as well as the left and right cursor to move to the points of intersection. Notice that the names of the graphs are shown in the upper left of the screen.

Section 5-3 Linear Programming in Two Dimensions: Geometric Approach

Example 2: Solving a Linear Programming Problem (page 280).
To solve this example, we will use the same procedure as we did in Example 2, section 5-2. You will use the application that graphs inequalities. Select the APPS key and scroll to the application **Inequalz**.

Solve each inequality for y and then enter into the Y=menu.

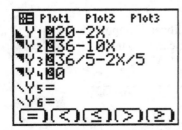 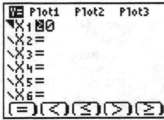 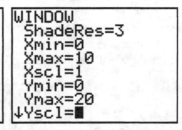

Graph the inequalities, then select the intersection by pressing ALPHA F1 and selecting option 1.
To find the points of intersection, press ALPHA F3 and move the cursor to the intersection points.

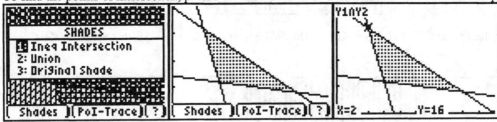

You can go to the home screen to evaluate the constraint equation.

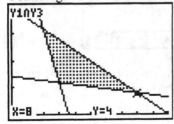 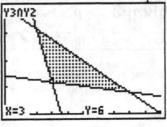 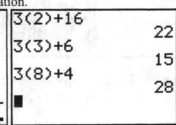

Chapter 6: Linear Programming – Simplex Method

For additional help on all aspects of the TI-83 or TI-84 graphing calculator, go to
http://www.prenhall.com/divisions/esm/app/graphing/ti83/

Section 6-1 A Geometric Introduction to the Simplex Method

No problems with the calculator in this section.

Section 6-2 The Simplex Method: Maximization with Problem Constraints of the Form $\leq$

Example 1: Using the Simplex Method (page 307).
Enter the simplex tableau in matrix [A]. Now perform the row operations to create the matrix that shows the optimal solution. The row operations are found in the matrix MATH menu.

```
[A]
[[4    1  1  0    2...
 [2    3  0  1  24  -...
 [-5   0  0  1  0    0...
```

```
NAMES MATH EDIT
0↑cumSum(
A:ref(
B:rref(
C:rowSwap(
D:row+(
E:*row(
F:*row+(
```

```
[2    3  0  1  24  -...
[-5   0  0  1  0    0...
*row(1/4,[A],1)→
[A]
[[1    .25 .25 0 ...
 [2    3   0   1 ...
 [-5   0   0   1 ...
```

```
*row+(-2,[A],1,2
)→[A]
[[1    .25 .25 0...
 [0    2.5 -.5 1...
 [-10 -5   0   0...
■
```

```
[0    2.5 -.5 1...
[-10 -5   0   0...
*row+(10,[A],1,3
)→[A]
[[1  .25   .25 0 ...
 [0  2.5  -.5  1 ...
 [0  -2.5 2.5  0 ...
■
```

```
[0 2.5   -.5 1 ...
[0 -2.5 2.5  0...
*row(1/25,[A],2)
→[A]
[[1 .25   .25   0...
 [0 .1   -.02  ....
 [0 -2.5 2.5   0...
```

```
*row(1/2.5,[A],2
)→[A]
[[1  .25   .25 0 ...
 [0  1    -.2  .4...
 [0 -2.5  2.5  0 ...
```

```
[0 1      -.2 .4...
[0 -2.5 2.5 0...
*row+(-.25,[A],2
,1)→[A]
[[1 0      .3   -...
 [0 1     -.2  .4...
 [0 -2.5  2.5  0 ...
■
```

```
[0 1      -.2 .4...
[0 -2.5 2.5 0...
*row+(2.5,[A],2,
3)→[A]
[[1 0  .3  -.1 0...
 [0 1 -.2  .4  0...
 [0 0  2    1   1...
■
```

Once you have performed all of the row operations, you can use the cursor to scroll to the right to see the solutions.

```
[0 1      -.2 .4...
[0 -2.5 2.5 0 ...
*row+(2.5,[A],2,
3)→[A]
...  .3   -.1 0  6 ]
...  -.2  .4  0  4 ]
...   2    1   1 80]]
■
```

Section 6-3 Dual Problem: Minimization with Problem Constraints of the Form ≥

Example 1: Forming the Dual Problem (page 317)
Enter matrix [A] and then create the transpose of the matrix.

```
[A]                   NAMES MATH EDIT        [4   1   1   30]
 [[2   1   5   20]    1:det(                 [40 12  40   1 ]]
  [4   1   1   30]    2:■ᵀ                [A]ᵀ
  [40 12 40   1 ]]    3:dim(
 ■                    4:Fill(                  [[2    4   40]
                      5:identity(              [1    1   12]
                      6:randM(                 [5    1   40]
                      7↓augment(               [20  30   1 ]]
                                             ■
```

Example 2: Solving a Minimization Problem (page 320).
Once you have created the tableau using the transpose matrix from Example 1, you follow the same procedures as in section 6-2, example 1, shown on the previous page.

Section 6-4 A Maximization and Minimization with Mixed Problem Constraints

No new methods of solutions are done in this section. You can use the methods covered in sections 6-2 and 6-3

Chapter 7: Logic, Sets, and Counting

For additional help on all aspects of the TI-83 or TI-84 graphing calculator, go to
http://www.prenhall.com/divisions/esm/app/graphing/ti83/

Section 7-1 Logic

To access the Logic menu, press 2nd Math, which is the TEST menu, then use the cursor to select the LOGIC menu. There are four logic operations that you can use. The operations **and, or,** and **xor** (exclusive or) return a value of 1 if an expression is true or 0 if an expression is false. The operation **not** returns 1 if the value is 0.

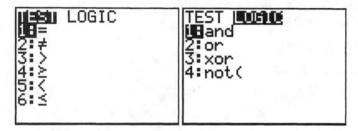

Example 5: Constructing a Truth Table (page 355) Construct the truth table for $p \wedge \neg(p \vee q)$.

Enter the values for p and q in L1 and L2. Press STAT and EDIT. Then you can put the values into the lists.

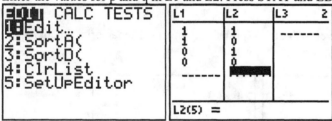

Now you will use the logic menu to put values into other lists. Press 2nd STAT (LIST) to access the names of the lists that are stored in the calculator. Then use the LOGIC menu as instructed above, and store the values in the given lists as shown.

Example 7: Verifying a Logical Equivalence (page 357) Show that $\neg(p \vee q) \equiv \neg p \wedge \neg q$.

Enter the values for p and q in L1 and L2 as shown in the previous example. Store $p \vee q$ in L3 and store $\neg(p \vee q)$ **in** L4. Store $\neg p \wedge \neg q$ in L5. You can see that L4 and L5 are equivalent.

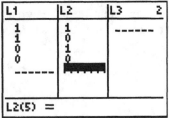 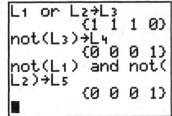

Section 7-2 Sets and Section 7-3

No problems from 7-2 or 7-3 can be done on the calculator.

Section 7-4 Permutations and Combinations

Example 1: Computing Factorials (page 375)
To calculate a factorial, enter MATH and scroll over to the PRB (probability) menu. The factorial function is selection 4. Now you can calculate the factorial expressions.

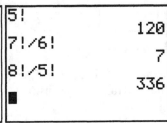

Example 3: Permutations (page 378) Find the number of permutations of 13 objects taken 8 at a time.
The permutation function is found on the probability menu as in the last example. It is the 2nd choice on the menu. You must enter the 13, then access **nPr** and then enter the 8.

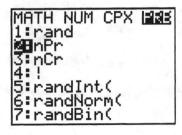 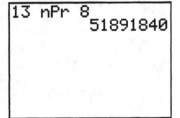

Example 5: Combinations (page 381). Find the number of combinations of 13 objects taken 8 at a time.
The combination function is found on the probability menu as in the last example. It is the 3rd choice on the menu. You must enter the 13, then access **nCr** and then enter the 8.

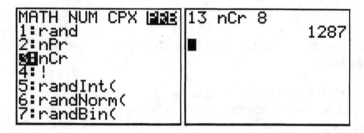

Chapter 8: Probability

For additional help on all aspects of the TI-83 or TI-84 graphing calculator, go to
http://www.prenhall.com/divisions/esm/app/graphing/ti83/

Section 8-1 Sample Spaces, Events, and Probability

Example 6: Simulation and Empirical Probabilities (page 400) Simulate 100 rolls of two dice with the random number generator on a graphing calculator.

The random number generator is found in the MATH PRB menu. This will generate integers. What we are going to do is have the calculator generate 100 random integers between 1 and 6, and this will be done twice. These two random numbers will be added together and stored in a list. We will use L1. Then we will create a histogram of the values found. Note that the same numbers will not be generated by your calculator that are generated by the example shown in the text. Recall that the list names are found by pressing 2nd STAT.

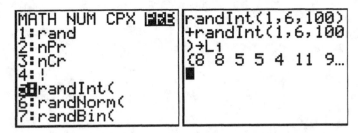

Now we will create a histogram. Press 2nd Y= to call up the STAT PLOT menu. Press ENTER and turn the plot on. Select the histogram icon. Set up a window.

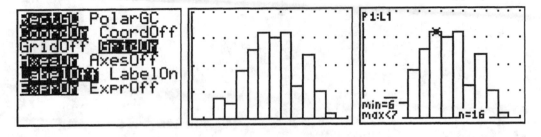

Press 2nd ZOOM (FORMAT) and select GridOn. Then press GRAPH. To see the values, use the TRACE key.

Section 8-2 Union, Intersection, and Complement of Events; Odds

Exercise 69, page (417). Use a graphing calculator to simulate 50 repetitions of rolling a pair of dice and recording their sum, and find the empirical probability of rolling a 7 or 8.

We will use the same commands as in the previous section. The only change will be that the number of repetitions will be changed to 200. Then the histogram will be created, and the TRACE key will be used to find the number of times that a sum of 7 occurred and the number of times that a sum of 8 occurred.

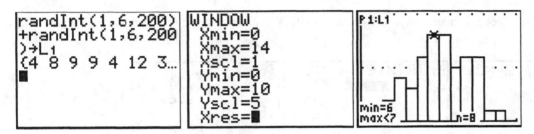

Both sums occurred 8 times, so the probability is (8+8)/50=0.32.

Section 8-3 Conditional Probability, Intersection, and Independence

No problems in this section introduce any new concepts. For directions on how to use the calculator to find factorials, combinations and permutations, see sections 7-3, and 7-4.

Section 8-4 Bayes' Formula

No problems in this section introduce any new concepts. For directions on how to use the calculator to find factorials, combinations and permutations, see sections 7-3, and 7-4.

Section 8-5 Random Variable, Probability Distribution, and Expected Value

No problems in this section introduce any new concepts. For directions on how to use the calculator to generate a table of random values, see the examples in section 8-1 and 8-2.

Chapter 9: Markov Chains

For additional help on all aspects of the TI-83 or TI-84 graphing calculator, go to
http://www.prenhall.com/divisions/esm/app/graphing/ti83/

Section 9-1 Properties of Markov Chains

Example 3: Using a Graphing Calculator and P^k to Compute S_k (page 460).

Store the matrices. Place P in matrix [A] and S_0 in matrix [B]. You can use the home screen to check your values.

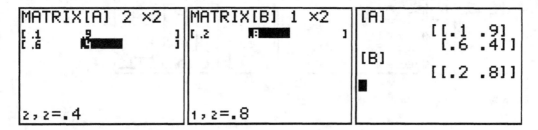

Now calculate $S_8 = S_0 P^8$.

```
[B]*[A]^8
[[.39921875 .60…
■
```

Section 9-2 Regular Markov Chains

Example 5: Approximating the Stationary Matrix (page 471). Use a graphing calculator to find a stationary matrix.

Put the transition matrix into the calculator. Set the decimal display to four places using the MODE menu. Compute powers of the matrix until all the rows are the same.

Scroll to the right to see the last column.

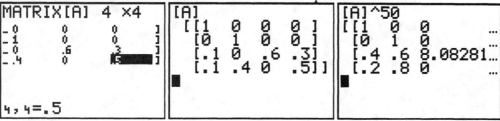

We can conclude that the transition matrix is

S=[.4943   .1494   .3563]

Section 9-3 Absorbing Markov Chains

Example 5: University Enrollment (page 484) Use a graphing calculator to find the limiting matrix.

Store the data in a matrix. Then raise the matrix to the 50th power.

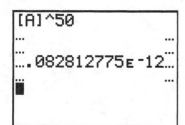

Scrolling to the right you see that the component of row 3 and column 3 is essentially a 0.

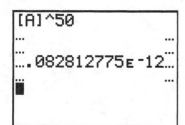

The matrix shows that the limiting matrix is $\begin{bmatrix} .4 & .6 \\ .2 & .8 \end{bmatrix}$.

From this you can determine that in the long run 60% of the first-year students will graduate and 20% of the second-year students will not graduate.

GETTING STARTED WITH MICROSOFT EXCEL®

DALE R. BUSKE *St. Cloud State University*

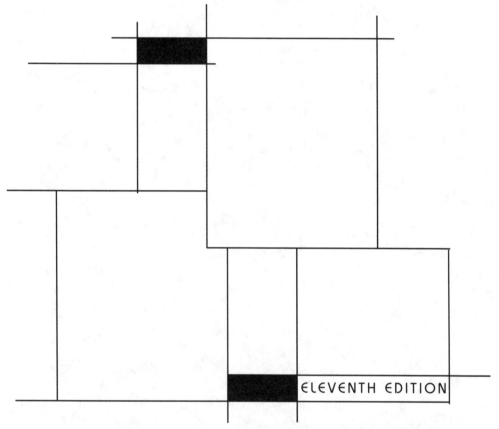

ELEVENTH EDITION

FINITE MATHEMATICS

For Business, Economics, Life Sciences, and Social Sciences

RAYMOND A. BARNETT
MICHAEL R. ZIEGLER
KARL E. BYLEEN

PEARSON

Prentice Hall

Upper Saddle River, NJ 07458

Table of Contents

Chapter 1

Section 1-1 Linear Equations and Inequalities

Example 1: Solving a Linear Equation

While linear equations can be "solved" using Excel (see Example 2), for now we focus on checking solutions to algebraic equations.

Enter: -9 in cell **A2**.

Enter: Formula **=8*A2-3*(A2-4) in cell B2.**

	A	B	C
1	x	LHS	RHS
2	-9	=8*A2-3*(A2-4)	

Enter: Formula **=3*(A2-4)+6 in cell C2.**

	A	B	C	D
1	x	LHS	RHS	
2	-9	-33	=3*(A2-4)+6	

Since the resulting values in cells **B2** and **C2** match, $x = -9$ is a solution.

	A	B	C
1	x	LHS	RHS
2	-9	-33	-33

Example 2: Solving a Linear Equation

To solve $\dfrac{x+2}{2} - \dfrac{x}{3} = 5$ for x,

Enter: **Any value (e.g. 0) in cell A2.**

Enter: Formula **=(A2+2)/2 – A2/3 in cell B2.**

	A	B	C
1	x	LHS	
2	0	=(A2+2)/2 - A2/3	

Choose: **Tools > Goal Seek**

Enter: Set cell: **B2**

Enter: To value: **5**

Enter: By changing cell: **A2**

Click: **OK**

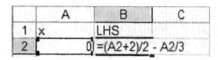

1

Goal Seek Status window will appear.

Click: **OK**

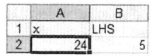

A "solution" will appear in cell **A2.** In general, when solving linear equations using **Goal Seek** in Excel the solution that appears will not be exact.

Example 4: Inequalities

Enter: **=3<5 in cell A1.**

	A
1	=3<5

Since the inequality is TRUE, the value **TRUE** appears in cell **A1**.

	A
1	TRUE

Example 8: Purchase Price

Enter: **Any value (e.g. 1) for the unknown price** x **in cell A2.**

Enter: **The value 57 in cell B2.**

Enter: **Formula =0.052*A2 in cell C2.**

Enter: **Formula =SUM(A2:C2) in cell D2.**

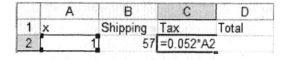

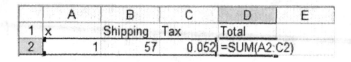

Choose: **Tools > Goal Seek**

Enter: Set cell: **D2**

Enter: To value: **851.26**

Enter: By changing cell: **A2**

Click: **OK**

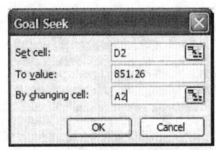

Goal Seek Status window will appear.

Click: **OK**

The solution will appear in cell **A2**.

	A	B	C	D
1	x	Shipping	Tax	Total
2	755	57	39.26	851.26

Example 9: Break-Even Analysis

Enter: **Any value (e.g. 1) for the unknown number of CD's *x* in cell A2.**

Enter: **Value 24000 in cell B2.**

Enter: **Formula =6.20*A2 in cell C2.**

	A	B	C	D	E
1	x	Fixed Costs	Variable Costs	Total Costs	Revenue
2	1	24000	=6.20*A2		

Enter: **Formula =SUM(B2:C2) in cell D2.**

	A	B	C	D	E
1	x	Fixed Costs	Variable Costs	Total Costs	Revenue
2	1	24000	6.2	=SUM(B2:C2)	

Enter: **Formula =8.70*A2 in cell E2.**

	A	B	C	D	E
1	x	Fixed Costs	Variable Costs	Total Costs	Revenue
2	1	24000	6.2	24006.2	=8.70*A2

Enter: **Formula =E2-D2**
in cell F2.

	A	B	C	D	E	F
1	x	Fixed Costs	Variable Costs	Total Costs	Revenue	Revenue-Costs
2	1	24000	6.2	24006.2	8.7	=E2-D2

Choose: **Tools > Goal**
Seek

Enter: Set cell: **F2**

Enter: To value: **0**

Enter: By changing cell:
A2

Click: **OK**

Goal Seek Status window
will appear.

Click: **OK**

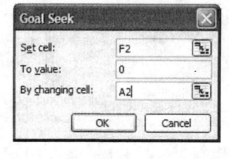

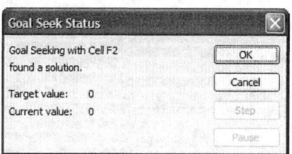

The solution will appear in
cell **A2.**

	A	B	C	D	E	F
1	x	Fixed Costs	Variable Costs	Total Costs	Revenue	Revenue-Costs
2	9600	24000	59520	83520	83520	0

Example 10: Consumer Price Index

Enter: **Any value for the**
unknown salary *x* **in cell**
C3 (e.g. 15000).

	A	B	C
1		1960	2000
2	CPI	29.6	172.2
3	Salary	13000	

Enter: **Formula =C2/B2**
in cell E2.

	A	B	C	D	E
1		1960	2000		Ratio
2	CPI	29.6	172.2		=C2/B2
3	Salary	13000	15000		

Enter: **Formula =C3/B3 in cell E3.**

	A	B	C	D	E
1		1960	2000		Ratio
2	CPI	29.6	172.2		5.817568
3	Salary	13000	15000		=C3/B3

Enter: **Formula =E3-E2 in cell F3.** [This is the difference of the ratios which we will set to 0.]

	A	B	C	D	E	F
1		1960	2000		Ratio	Difference
2	CPI	29.6	172.2		5.817568	
3	Salary	13000	15000		1.153846	=E3-E2

Choose: **Tools > Goal Seek**

Enter: Set cell: **F3**

Enter: To value: **0**

Enter: By changing cell: **C3**

Click: **OK**

Goal Seek Status window will appear.

Click: **OK**

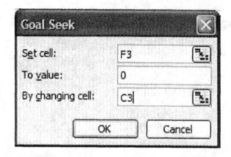

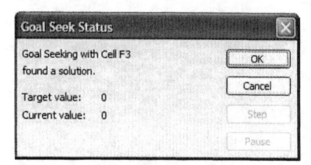

The solution will appear in cell **C3.**

	A	B	C	D	E	F
1		1960	2000		Ratio	Difference
2	CPI	29.6	172.2		5.817568	
3	Salary	13000	75628.38		5.817568	0

Section 1-2 Graphs and Lines

Example 2: Using a Graphing Calculator (or Excel!)

Enter *x*-values starting at -10 and increasing by 1.

Enter: **The value -10 in cell A2.**

Enter: **Formula =A2+1 in cell A3.**

Copy: **Contents of cell A3 to cells A4:A22.**

Enter: **Formula =3/4*A2-3 in cell B2.**

Copy: **Contents of cell B2 to cells B3:B22.**

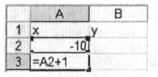

Choose: **Chart Wizard >
(XY) Scatter >**

Select: **Chart sub-type:
Scatter with data points
connected by smoothed
lines.**

Click: **Next**

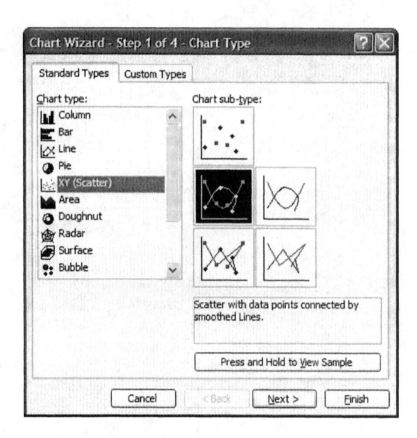

Enter: **Data Range**
A2:B22 or select cells

Select: **Series in:**
columns

Click: **Next**

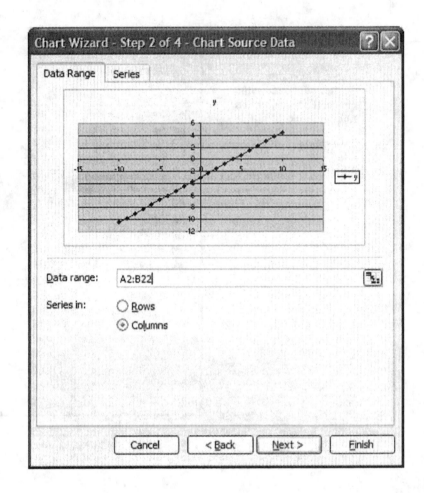

Enter: **Chart title, Value (X) axis, and Value (Y) axis**

Click: **Finish**

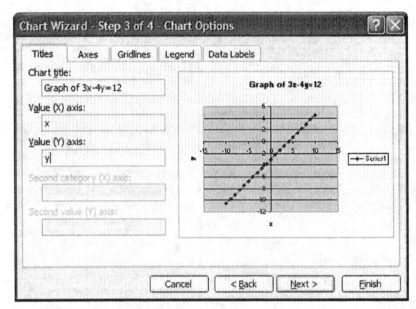

Right Click: **On x-axis.**

Select: **Format Axis**

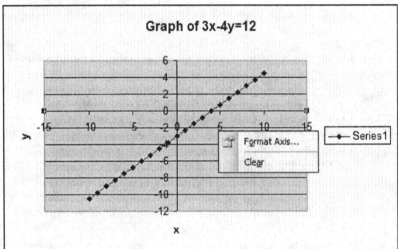

Select: **Scale Tab**

Enter: **Minimum Value of -10**

Enter: **Maximum Value of 10**

Click: **OK**

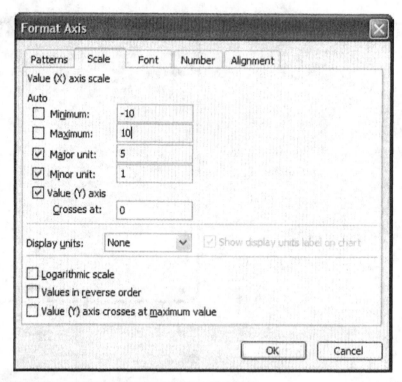

Right Click: **On *y*-axis.**

Select: **Format Axis**

Select: **Scale Tab**

Enter: **Minimum Value of -5**

Enter: **Maximum Value of 5**

Click: **OK**

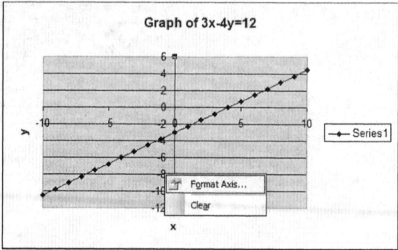

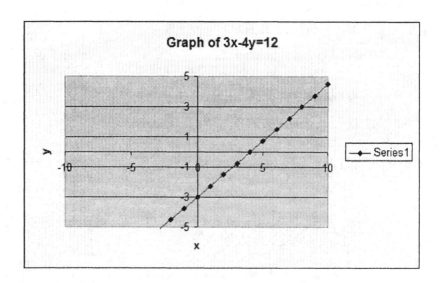

Example 4: Finding Slopes

Enter: **The value -3 in cell A2.**

Enter: **The value 3 in cell A3.**

Enter: **The value -2 in cell B2.**

Enter: **The value 4 in cell B3.**

	A	B	C	D	E
1	x	y		Slope	
2		-3	-2	=(B3-B2)/(A3-A2)	
3		3	4		

Enter: **Formula =(B3-B2)/(A3-A2)
in cell D2.**

	A	B	C	D
1	x	y		Slope
2		-3	-2	1
3		3	4	

To compute the slope of a new line, simply change the values of the ordered pairs in cells **A2, A 3, B 2**, and **B3**.

Enter: **The value -1 in cell A2.**

Enter: **The value 3 in cell A3.**

Enter: **The value 2 in cell B2.**

Enter: **The value -3 in cell B3.**

	A	B	C	D
1	x	y		Slope
2		-1	2	-1.25
3		3	-3	

Example 8: Supply and Demand

Copy the data from the text (Table 3) into the worksheet.

	A	B	C	D
1	Year	Supply (mil bu)	Demand (mil bu)	Price ($/bu)
2	1997	2480	2300	3.38
3	1999	2300	2390	2.48
4				
5		Slope (Supply)	Slope (Demand)	
6		=(D2-D3)/(B2-B3)		

To find the slope of the supply equation,

Enter: **Formula =(D2-D3)/(B2-B3) in cell B6.**

To find the slope of the demand equation,

Enter: **Formula =(D2-D3)/(C2-C3) in cell C6.**

	A	B	C	D
1	Year	Supply (mil bu)	Demand (mil bu)	Price ($/bu)
2	1997	2480	2300	3.38
3	1999	2300	2390	2.48
4				
5		Slope (Supply)	Slope (Demand)	
6		0.005	=(D2-D3)/(C2-C3)	

Enter: **Formula =B6*(A9-B3)+D3 in cell B9.**

	A	B	C	D
1	Year	Supply (mil bu)	Demand (mil bu)	Price ($/bu)
2	1997	2480	2300	3.38
3	1999	2300	2390	2.48
4				
5		Slope (Supply)	Slope (Demand)	
6		0.005	-0.01	
7				
8	x (Quantity)	Supply	Demand	
9	2300	=B6*(A9-B3)+D3		

Enter: **Formula =C6*(A9-C3)+D3 in cell C9.**

	A	B	C	D
1	Year	Supply (mil bu)	Demand (mil bu)	Price ($/bu)
2	1997	2480	2300	3.38
3	1999	2300	2390	2.48
4				
5		Slope (Supply)	Slope (Demand)	
6		0.005	-0.01	
7				
8	x (Quantity)	Supply	Demand	
9	2300	2.48	=C6*(A9-C3)+D3	

Enter: **Formula =B9-C9 in cell D9.**

Our goal will be to determine for what quantity (x) this value (the difference between the supply and demand) is equal to 0.

	A	B	C	D
1	Year	Supply (mil bu)	Demand (mil bu)	Price ($/bu)
2	1997	2480	2300	3.38
3	1999	2300	2390	2.48
4				
5		Slope (Supply)	Slope (Demand)	
6		0.005	-0.01	
7				
8	x (Quantity)	Supply	Demand	Difference
9	2300	2.48	3.38	=B9-C9

Choose: **Tools > Goal Seek**

Enter: Set cell: **D9**

Enter: To value: **0**

Enter: By changing cell: **A9**

Click: **OK**

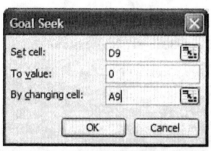

Goal Seek Status window will appear.

Click: **OK**

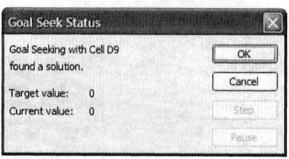

The equilibrium quantity will appear in cell **A9**. The equilibrium price will appear in cells **B9** and **C9**.

	A	B	C	D
1	Year	Supply (mil bu)	Demand (mil bu)	Price ($/bu)
2	1997	2480	2300	3.38
3	1999	2300	2390	2.48
4				
5		Slope (Supply)	Slope (Demand)	
6		0.005	-0.01	
7				
8	x (Quantity)	Supply	Demand	Difference
9	2360	2.78	2.78	0

Section 1-3 Linear Regression

Example 3: Diamond Prices

Copy the data from
the text into the
worksheet.

	A	B
1	Weight	Price
2	0.5	2790
3	0.6	3191
4	0.7	3694
5	0.8	4154
6	0.9	5018
7	1	5898

Choose: **Chart
Wizard > XY
(Scatter) > Next**

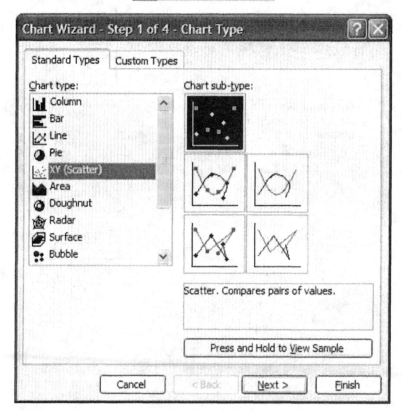

Enter: **Data Range
A2:B7** or select cells

Select: **Series in:
columns**

Click: **Next**

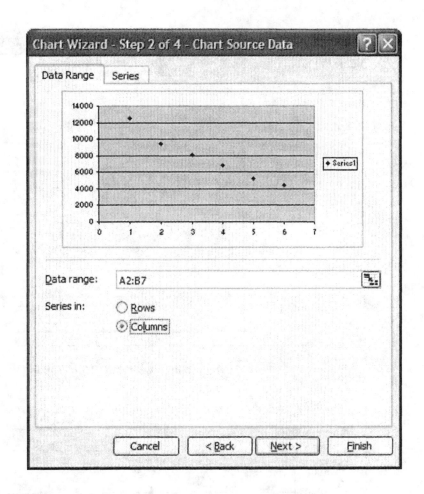

Enter: **Chart title,
Value (X) axis, and
Value (Y) axis**

Click: **Finish**

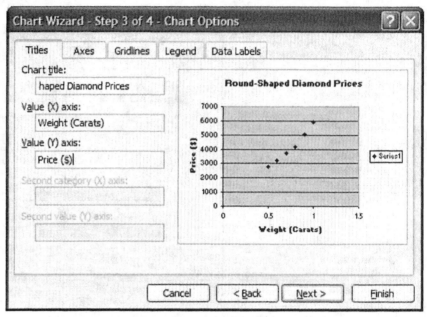

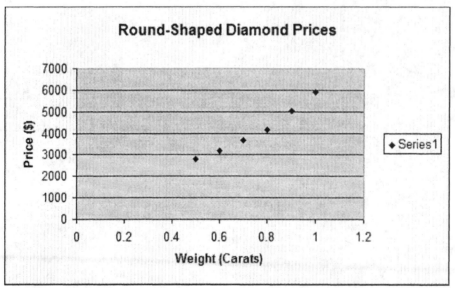

Choose: **Chart > Add Trendline**

Select: **Linear**

Select: **Options Tab**

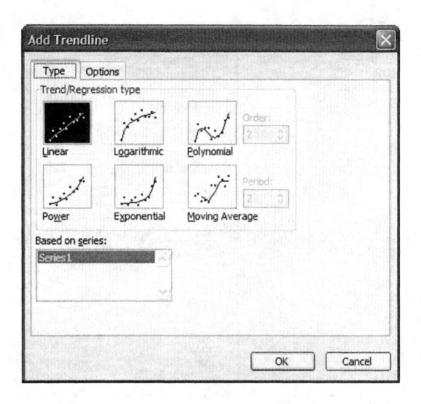

Select: **Display equation on chart**

Click: **OK**

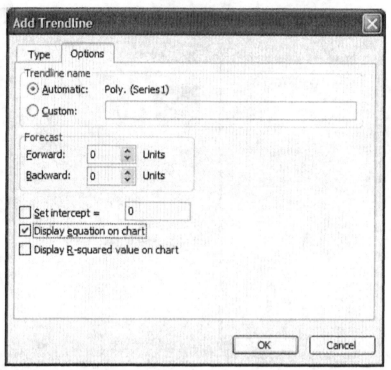

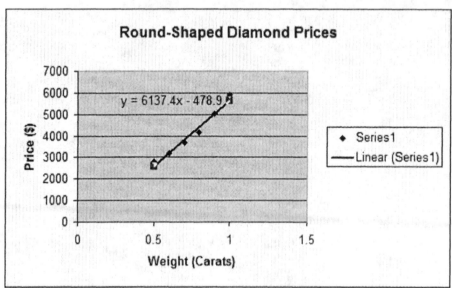

To estimate the value of 0.85 and 1.2 carat diamonds,

Enter: **Value 0.85 in cell A9**

Enter: **Value 1.2 in cell A10**

Enter: **Formula =6137.4*A9-478.9 in cell B9**

	A	B	C
1	Weight	Price	
2	0.5	2790	
3	0.6	3191	
4	0.7	3694	
5	0.8	4154	
6	0.9	5018	
7	1	5898	
8			
9	0.85	=6137.4*A9-478.9	
10	1.2		

Copy: **The contents of cell B9 to cell B10** [Highlight cell B9, place cursor at lower left corner, hold left mouse button down and drag down]

	A	B
1	Weight	Price
2	0.5	2790
3	0.6	3191
4	0.7	3694
5	0.8	4154
6	0.9	5018
7	1	5898
8		
9	0.85	4737.89
10	1.2	6885.98

To estimate the weight of a $4,000 diamond,

Enter: *Any* **guess (e.g. 0.75) to the value of such a diamond in cell A11**

Copy: **The contents of cell B10 to cell B11**

	A	B
1	Weight	Price
2	0.5	2790
3	0.6	3191
4	0.7	3694
5	0.8	4154
6	0.9	5018
7	1	5898
8		
9	0.85	4737.89
10	1.2	6885.98
11	0.75	4124.15

Choose: **Tools > Goal Seek**

Enter: Set cell: **B11**

Enter: To value: **4000**

Enter: By changing cell: **A11**

Click: **OK**

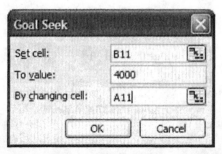

Goal Seek Status window will appear.

Click: **OK**

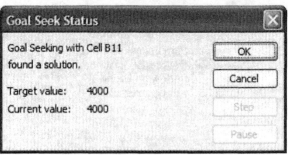

The estimated weight of a $4000 diamond (in Carats) will appear in cell **A11.**

	A	B
1	Weight	Price
2	0.5	2790
3	0.6	3191
4	0.7	3694
5	0.8	4154
6	0.9	5018
7	1	5898
8		
9	0.85	4737.89
10	1.2	6885.98
11	0.729772	4000

Chapter 2

Section 2-1 Functions

Example 1: Point-by-Point Plotting

Enter: **-4 in cell A2.**

Enter: **Formula =A2+1 in cell A3.**

	A	B
1	x	y
2	-4	
3	=A2+1	

Copy: **Contents of cell A3 to cells A4:A10.**

	A	B
1	x	y
2	-4	
3	-3	
4	-2	
5	-1	
6	0	
7	1	
8	2	
9	3	
10	4	

Enter: **Formula =9-A2^2 in cell B2.**

	A	B
1	x	y
2	-4	=9-A2^2
3	-3	
4	-2	
5	-1	
6	0	
7	1	
8	2	
9	3	
10	4	

Copy: **Contents of cell A3 to cells A4:A10.**

	A	B
1	x	y
2	-4	-7
3	-3	0
4	-2	5
5	-1	8
6	0	9
7	1	8
8	2	5
9	3	0
10	4	-7

Choose: **Chart Wizard > (XY) Scatter >**

Select: **Chart sub-type: Scatter with data points connected by smoothed lines.**

Click: **Next**

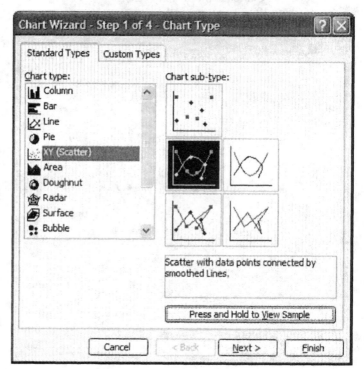

Enter: **Data Range A2:B10 or select cells**

Select: **Series in: columns**

Click: **Next**

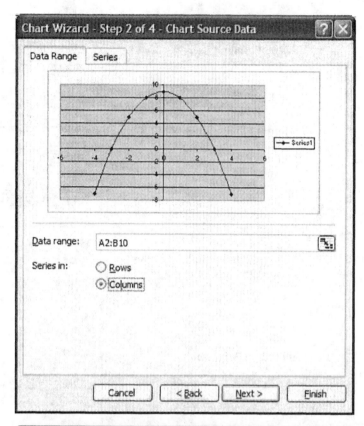

Select: **Gridlines Tab**

Select: **Value (X) axis: Major axis and Minor axis**

Select: **Value (Y) axis: Major axis and Minor axis**

Click: **Finish**

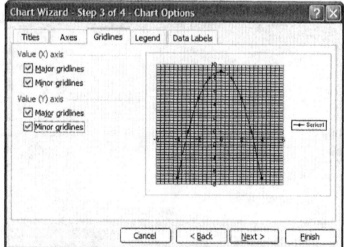

Right-Click: **On *x*-axis.**

Select: **Format Axis...**

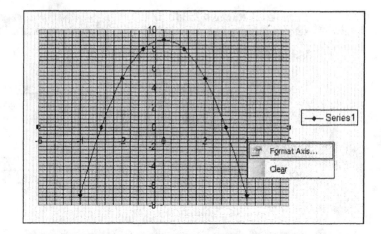

Select: **Scale Tab**

Enter: **-10 for Minimum**

Enter: **10 for Maximum**

Enter: **5 for Major unit**

Enter: **1 for Minor unit**

Click: **OK**

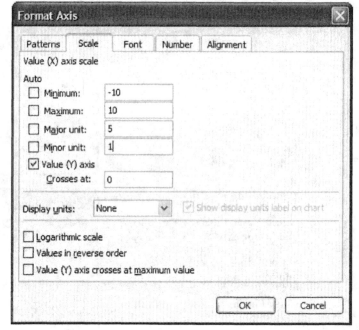

Right-Click: **On *y*-axis.**

Select: **Format Axis…**

Select: **Scale Tab**

Enter: **-10 for Minimum**

Enter: **10 for Maximum**

Enter: **5 for Major unit**

Enter: **1 for Minor unit**

Click: **OK**

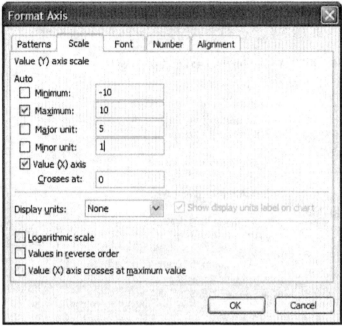

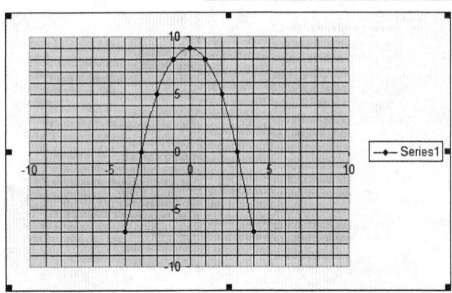

Example 4: Function Evaluation

Enter: **Value 6 in cell A2.**

Enter: **Formula =12/(A2-2) in cell B2.**

	A	B
1	x	f(x)
2	6	=12/(A2-2)

Note: The function value in cell **B2** will automatically update when new values are entered into cell **A2**.

	A	B
1	x	f(x)
2	6	3

Enter: **Value -2 in cell A2.**

Enter: **Formula =SQRT(A2-1) in cell B2.**

	A	B	C
1	x	h(x)	
2	-2	=SQRT(A2-1)	

Since -3 is not in the domain of the root function, an error message occurs in cell **B2**.

	A	B
1	x	h(x)
2	-2	#NUM!

Example 7: Price-Demand and Revenue Modeling

Copy the data from the text into the worksheet.

	A	B
1	x (Millions)	p ($)
2	2	87
3	5	68
4	8	53
5	12	37

Choose: **Chart Wizard > (XY) Scatter >**

Select: **Chart sub-type: Scatter**

Click: **Next**

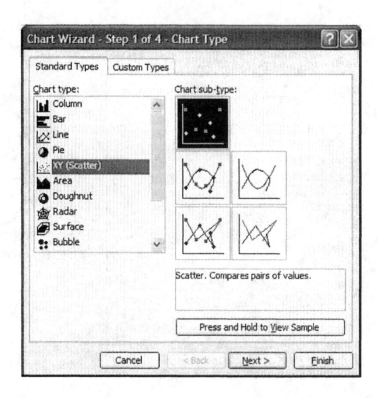

Enter: **Data Range A2:B5 or select cells**

Select: **Series in: columns**

Click: **Next**

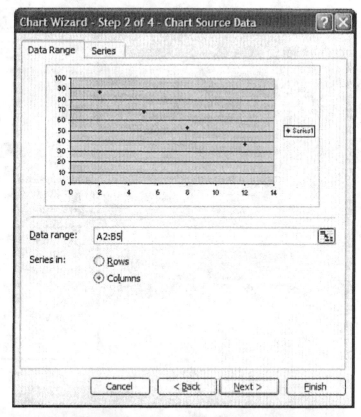

Select: **Titles Tab**

Enter: **Chart title, Value (X) axis, and Value (Y) axis**

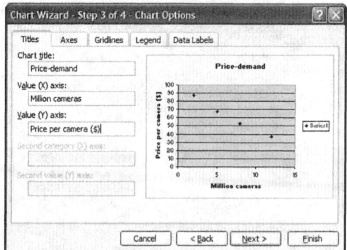

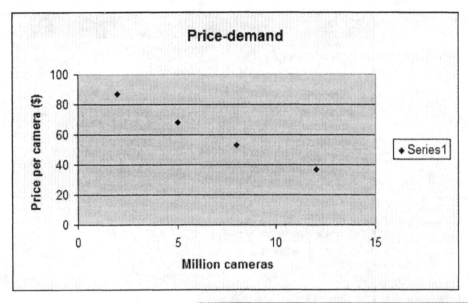

Choose: **Chart > Add Trendline**

Select: **Linear**

Select: **Options Tab**

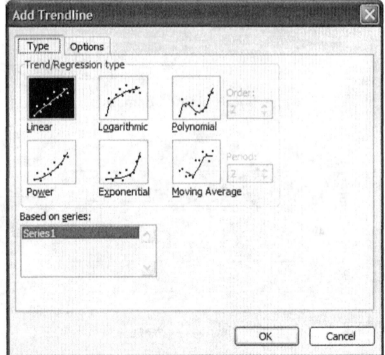

Select: **Display equation on chart**

Click: **OK**

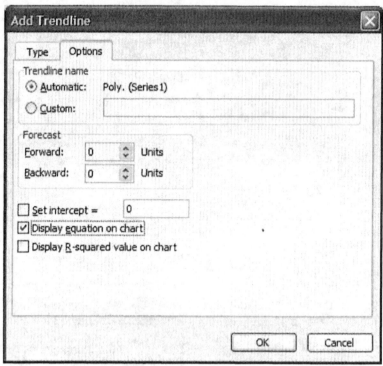

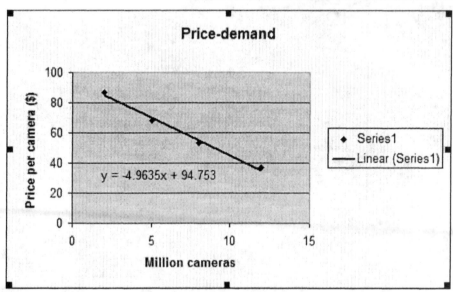

Enter: **Formula =A2*(-5*A2+94.8) in cell B2**

	A	B	C
1	x (Millions)	R(x) (Million $)	
2	1	=A2*(-5*A2+94.8	
3	3		
4	6		
5	9		
6	12		
7	15		

Copy: **The contents of cell B2 to cells B3:B7**

	A	B
1	x (Millions)	R(x) (Million $)
2	1	89.8
3	3	239.4
4	6	388.8
5	9	448.2
6	12	417.6
7	15	297

Choose: **Chart Wizard > (XY) Scatter >**

Select: **Chart sub-type: Scatter with data points connected by smoothed lines.**

Click: **Next**

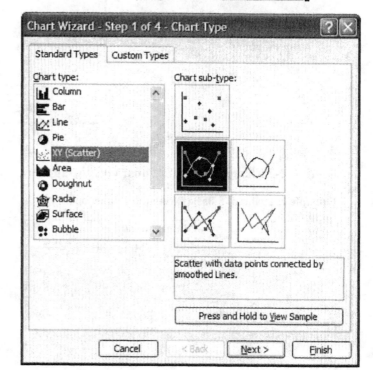

31

Enter: **Data Range A2:B7** or select cells

Select: **Series in: columns**

Click: **Finish**

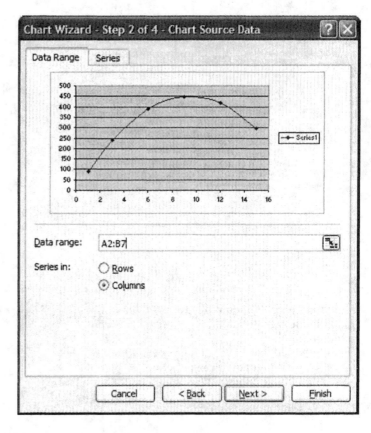

Section 2-2 Elementary Functions: Graphs and Transformations

Example 1: Evaluating Basic Elementary Functions

Mathematical Function	Excel Function
x^2	=x^2
x^3	=x^3
$\sqrt{x}$	=SQRT(x)
$\sqrt[3]{x}$	=x^(1/3)
$\lvert x \rvert$	=ABS(x)

Example 6: Natural Gas Rates

Piecewise defined functions can be implemented in Excel using the built-in IF function. The syntax of this function works as follows: IF(logical_test,value_if_true,value_if_false). In the following, Excel first checks whether the contents of cell **A2** has a value less than or equal to 0. If it does, then it outputs and error. Next, it checks if the contents have a value less than or equal to 5. Then, it checks to see if the contents are less than or equal to 40.

Enter: **Any value (e.g. 7) in cell A2.**

Enter: **Formula =IF(A2<0,'ERROR',IF(A2<=5,0.7866*A2,IF(A2<=40,3.933+0.4601*(A2-5),20.0365+0.2508*(A2-40)))) in cell B2.**

	A	B	C	D	E	F	G	H	I	J
1	x	C(x)								
2		7	=IF(A2<=0,"ERROR",IF(A2<=5,0.7866*A2,IF(A2<=40,3.933+0.4601*(A2-5),20.0365+0.2508*(A2-4							

Graphing a piecewise-defined function is no different than graphing any other function in Excel.

Enter: **0 in cell A2.**

Enter: **Formula =A2+1 in cell A3.**

	A
1	x
2	0
3	=A2+1

Copy: **Contents of cell A3 to cells A4:A62.**

	A	B
1	x	C(x)
2	0	0
3	1	
4	2	
5	3	
6	4	
7	5	

Enter: **Formula =IF(A2<0,'ERROR',IF(A2<=5,0.7 866*A2,IF(A2<=40,3.933+0.4601 *(A2-5),20.0365+0.2508*(A2- 40)))) in cell B2.** **Copy the contents of cell B2 to cells B3:B62.**

	A	B
1	x	C(x)
2	0	0
3	1	0.7866
4	2	1.5732
5	3	2.3598
6	4	3.1464
7	5	3.933
8	6	4.3931

Choose: **Chart Wizard > (XY) Scatter >**

Select: **Chart sub-type: Scatter with data points connected by smoothed lines.**

Click: **Next**

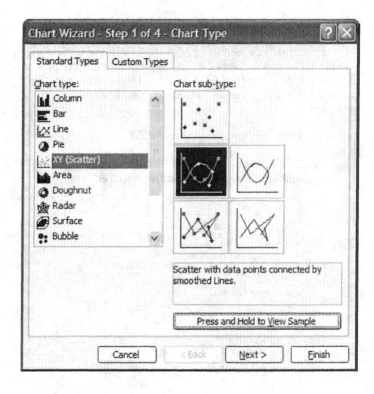

Enter: **Data Range A2:B62 or select cells**

Select: **Series in: columns**

Click: **Next**

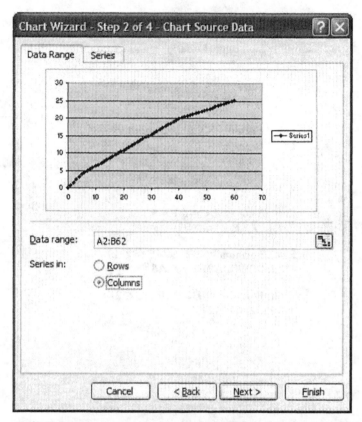

Select: **Titles Tab**

Enter: **Chart title, Value (X) axis, Value (Y) axis**

Click: **Finish**

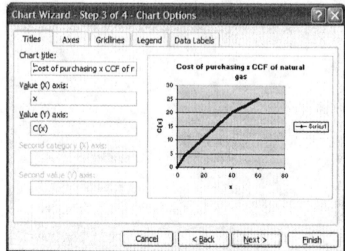

Section 2-3 Quadratic Functions

Example 1: Intercepts, Equations, and Inequalities

To find the *y*-intercept,

Enter: **0 in cell A2.**

	A	B	C
1	x	f(x)	
2	0	=-A2^2+5*A2+3	

Enter: **Formula =-A2^2+5*A2+3 in cell B2.**

To find the *x*-intercepts using the quadratic formula,

Enter: **Values of a = -1, b = 5, and c = 3 in cells A2,B 2, and C2**

	A	B	C	D
1	a	b	c	
2	-1	5	3	
3				
4	Root 1	Root 2		
5	=(-B2+SQRT(B2^2-4*A2*C2))/(2*A2)			

Enter: **Formula =(-B2+SQRT(B2^2-4*A2*C2))/(2*A2) in cell A5.**

Enter: **Formula =(-B2-SQRT(B2^2-4*A2*C2))/(2*A2) in cell B5.**

	A	B	C
1	a	b	c
2	-1	5	3
3			
4	Root 1	Root 2	
5	-0.54138	5.541381	

Example 2: Analyzing a Quadratic Function

Choose: **Tools > Add-ins >**

Select: **Solver Add-in**

Click: **OK**

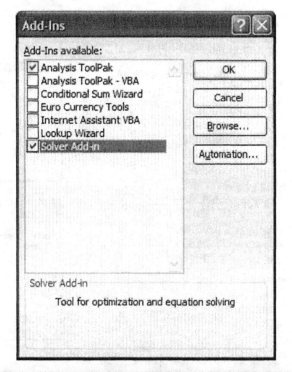

If this add-in has not been installed before, you will be prompted to install it now.

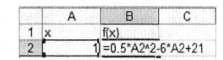

Enter: **Any value (e.g. 1) in cell A2.**

Enter: **Formula = 0.5*A2^2-6*A2+21 in cell B2.**

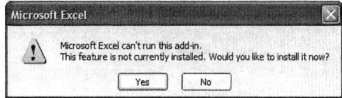

	A	B	C
1	x	f(x)	
2	1	=0.5*A2^2-6*A2+21	

Choose: **Tools > Solver >**

Enter: **Select Target Cell: B2**

Select: **Equal To: Min**

Enter: **By Changing Cells: A2**

Click: **Solve**

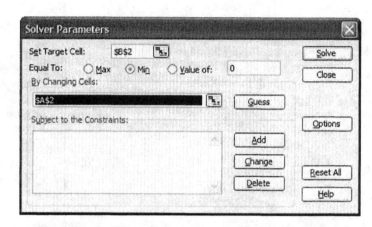

Select: **Keep Solver Solution**

Click: **OK**

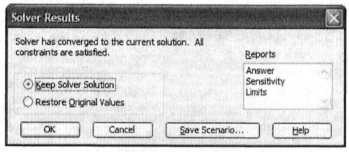

Since the parabola opens upward, the
coordinates for the vertex will appear
in cells **A2** and **B2**.

	A	B
1	x	f(x)
2	6	3

Example 3: Maximum Revenue

Choose: **Tools > Add-ins >**

Select: **Solver Add-in**

Click: **OK**

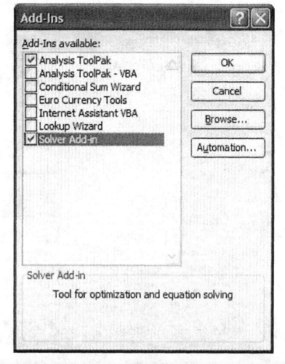

If this add-in has not been installed before, you will be prompted to install it now.

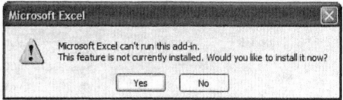

Enter: **Any value (e.g. 3) in cell A2.**

Enter: **Formula =A2*(94.8-5*A2) in cell B2.**

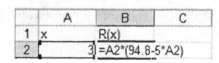

	A	B	C
1	x	R(x)	
2	3	=A2*(94.8-5*A2)	

Choose: **Tools > Solver >**

Enter: **Select Target Cell: B2**

Select: **Equal To: Max**

Enter: **By Changing Cells: A2**

Click: **Subject to the Constraints: Add**

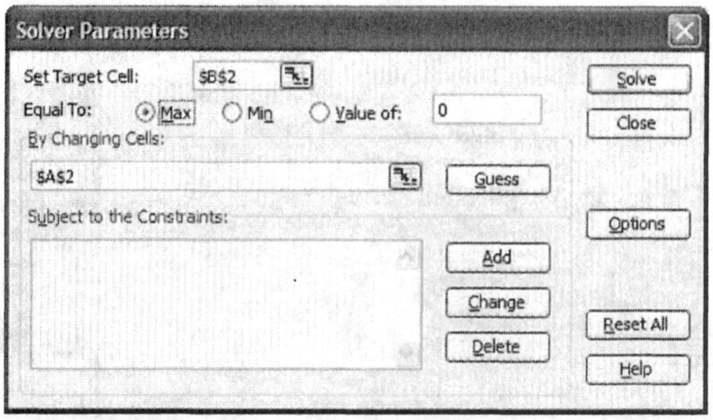

Enter: **Cell Reference: A2**

Select: **>=**

Enter: **Constraint 1**

Click: **Add**

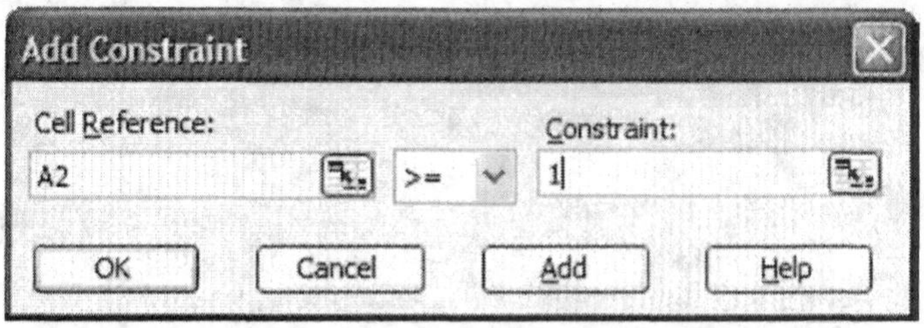

Enter: **Cell Reference: A2**

Select: **<=**

Enter: **Constraint 15**

Click: **OK**

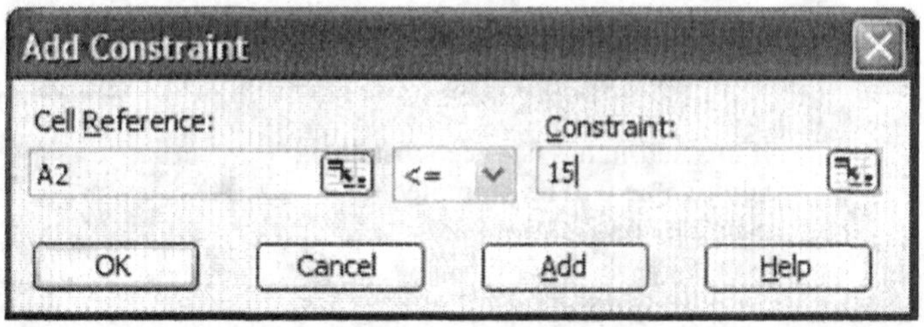

Click: **Solve**

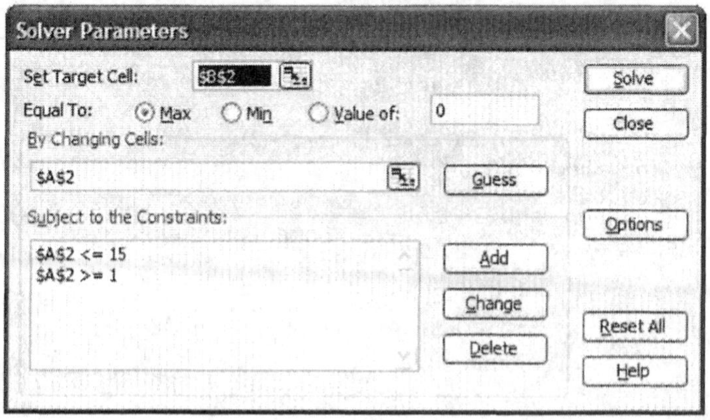

Select: **Keep Solver Solution**

Click: **OK**

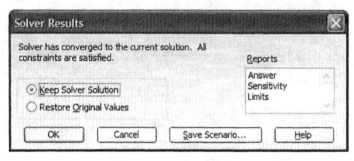

Since the parabola opens downward, the coordinates for the vertex of $R(x)$ will appear in cells **A2** and **B2**.

	A	B
1	x	R(x)
2	9.48	449.352

Example 4: Break-Even Analysis

Enter: **The value 1 in cell A2**

Enter: **Formula =A2+1 in cell A3**

	A	B	C
1	x	R(x)	C(x)
2		1	
3	=A2+1		

Copy: **Contents of cell A3** to cells **A4:A16**

Enter: **Formula =A2*(94.8-5*A2) in cell B2.**

	A	B	C
1	x	R(x)	C(x)
2		1	=A2*(94.8-5*A2)
3		2	
4		3	
5		4	
6		5	
7		6	
8		7	
9		8	
10		9	
11		10	
12		11	
13		12	
14		13	
15		14	
16		15	

Copy: **Contents of cell B2 to cells B3:B16**

Enter: **Formula =156+19.7*A2 in cell C2.**

Copy: **Contents of cell C2 to cells C3:C16**

	A	B	C	D
1	x	R(x)	C(x)	
2	1	89.8	=156+19.7*A2	
3	2	169.6		
4	3	239.4		
5	4	299.2		
6	5	349		
7	6	388.8		
8	7	418.6		
9	8	438.4		
10	9	448.2		
11	10	448		
12	11	437.8		
13	12	417.6		
14	13	387.4		
15	14	347.2		
16	15	297		

Choose: **Chart Wizard > (XY) Scatter >**

Select: **Chart sub-type: Scatter with data points connected by smoothed lines.**

Click: **Next**

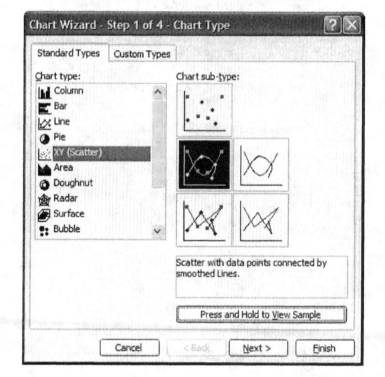

Enter: **Data Range A2:C6 or select cells**

Select: **Series in: columns**

Click: **Series Tab**

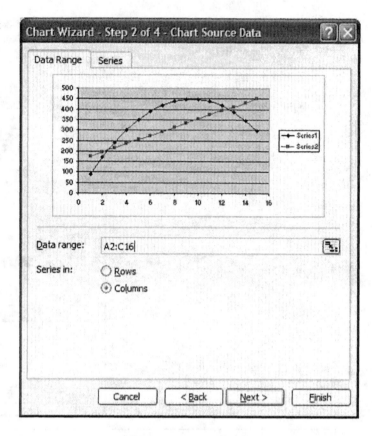

Select: **Series 1**

Enter: **Name: R(x)**

Select: **Series 2**

Enter: **Name: C(x)**

Click: **Finish**

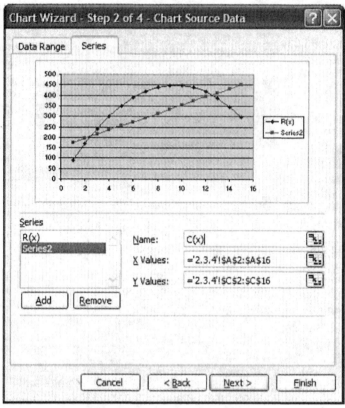

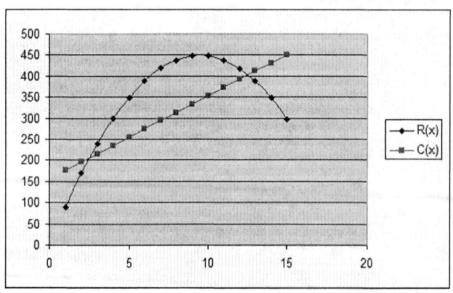

Enter: **Formula =B2-C2 in cell D2**

	A	B	C	D
1	x	R(x)	C(x)	
2	1	89.8	175.7	=B2-C2
3	2	169.6	195.4	
4	3	239.4	215.1	

Choose: **Tools > Goal Seek >**

Enter: **Set cell: D2**

Enter: **To value: 0**

Enter: **By changing cell: A2**

Click: **OK**

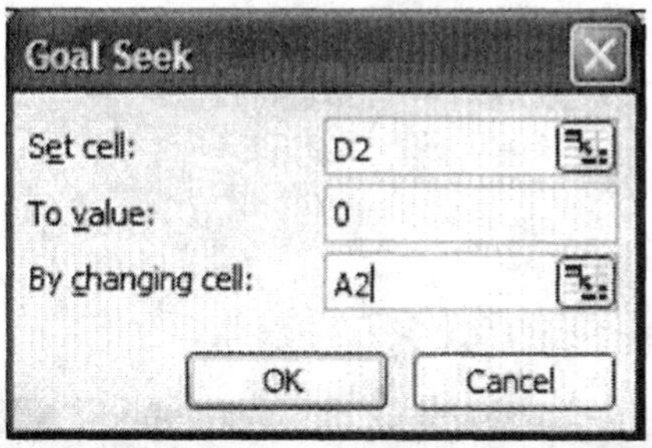

Goal Seek	
Set cell:	D2
To value:	0
By changing cell:	A2
	OK Cancel

Click: **OK**

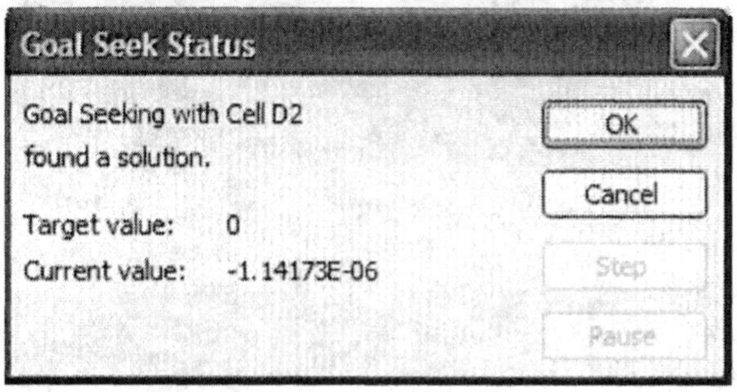

Goal Seek Status

Goal Seeking with Cell D2
found a solution.

Target value: 0
Current value: -1.14173E-06

OK
Cancel
Step
Pause

One of the two break-even points (the one closest to $x = 1$) will appear in cell **A2**.

	A	B	C	D
1	x	R(x)	C(x)	
2	2.49003	205.0536	205.0536	-1.1E-06

To find the other break-even point,

Enter: **A value close to the other break-even point in cell A2. (e.g. 12 or 13 found from the graph above)**

	A	B	C	D
1	x	R(x)	C(x)	
2	12.52997	402.8405	402.8404	0.000121

Choose: **Tools > Goal Seek >**

Enter: **Set cell: D2**

Enter: **To value: 0**

Enter: **By changing cell: A2**

Click: **OK**

Example 5: Outboard Motors

Copy the data from
the text into the
worksheet.

	A	B	C
1	RPM	MPH	MPG
2	2500	10.3	4.1
3	3000	18.3	5.6
4	3500	24.6	6.6
5	4000	29.1	6.4
6	4500	33	6.1
7	5000	36	5.4
8	5400	38.9	4.9

Choose: **Chart
Wizard > XY
(Scatter) > Next**

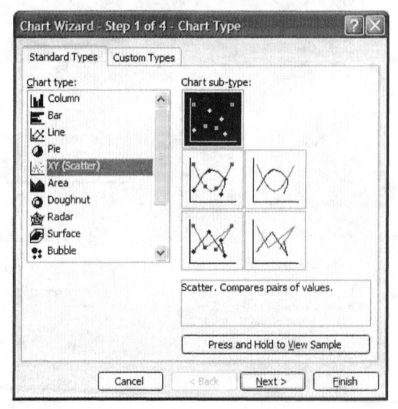

Enter: **Data Range
B2:C8** or select cells

Select: **Series in:
columns**

Click: **Next**

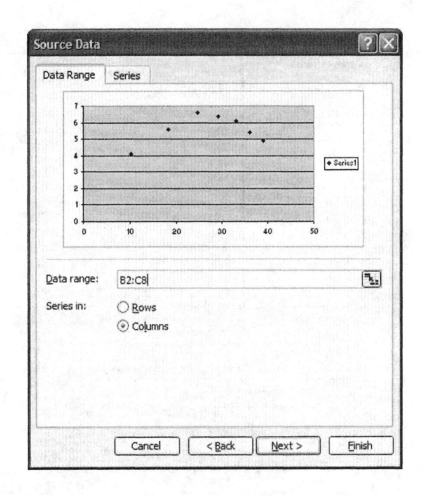

Enter: **Chart title,**
Value (X) axis, and
Value (Y) axis

Click: **Finish**

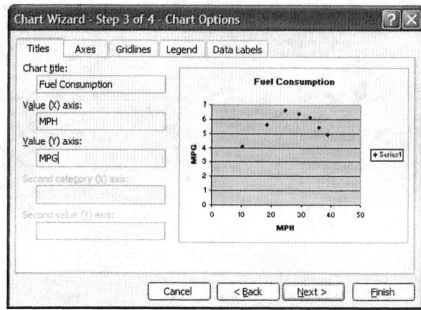

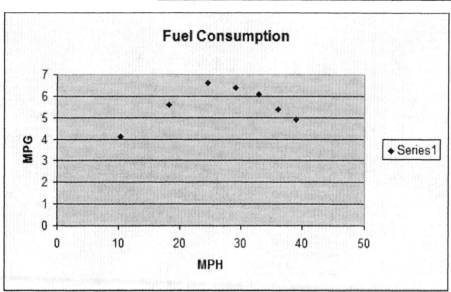

Click: **On the chart (highlight it)**

Choose: **Chart > Add Trendline**

Select: **Polynomial of Order 2**

Select: **Options Tab**

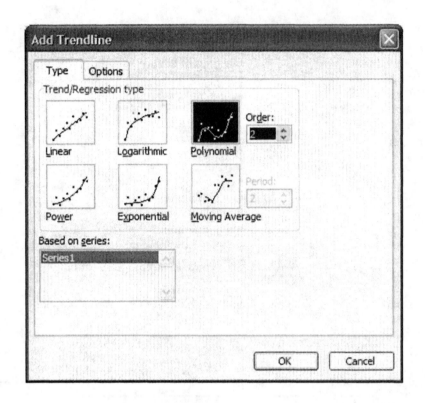

Select: **Display
equation on
chart**

Click: **OK**

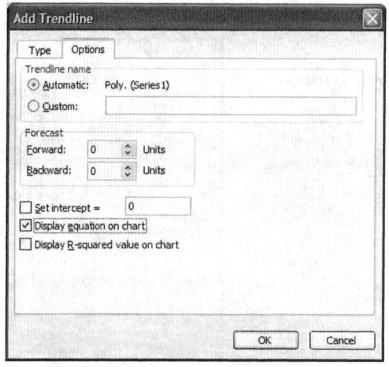

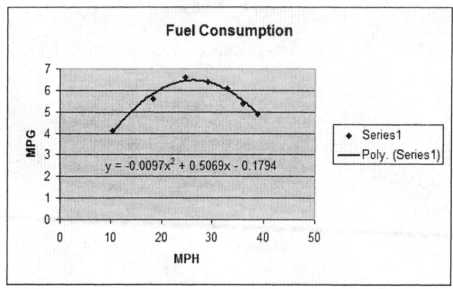

Section 2-4 Exponential Functions

Example 1: Graphing Exponential Functions

Enter *x*-values starting at -2 and
increasing by 1.

Enter: **The value -2 in cell A2.**

Enter: **Formula =A2+1 in cell A3.**

Copy: **Contents of cell A3 to cells
A4:A6.**

Enter: **Formula =(1/2)*4^A2 in cell
B2.**

Copy: **Contents of cell B2 to cells
B3:B6.**

	A	B
1	x	y
2		-2
3	=A2+1	

	A	B
1	x	y
2		-2	=(1/2)*4^A2
3		-1
4		0
5		1
6		2

Choose: **Chart Wizard >
(XY) Scatter >**

Select: **Chart sub-type:
Scatter with data points
connected by smoothed
lines.**

Click: **Next**

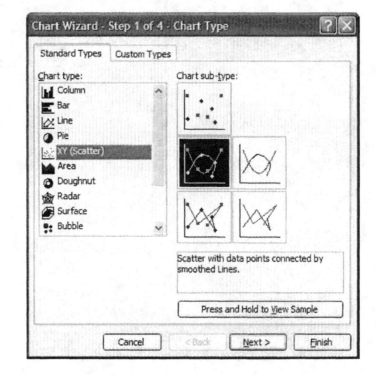

Enter: **Data Range
A2:B6** or select cells

Select: **Series in:
columns**

Click: **Next**

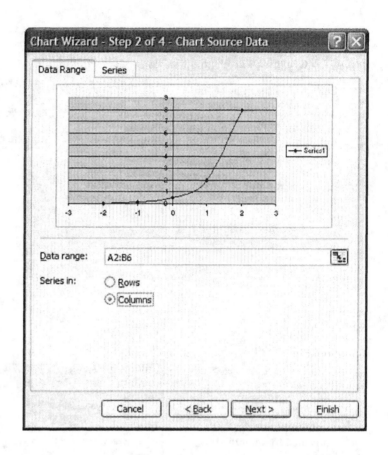

Enter: Chart title, Value (X) axis, and Value (Y) axis

Click: Finish

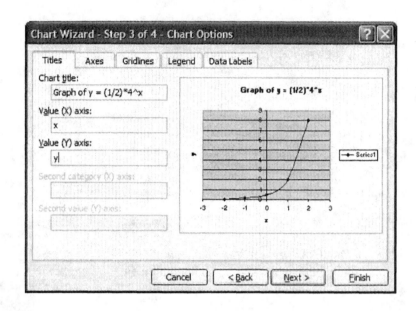

Example 2: Exponential Growth

Enter: **The value 0.6 in cell A2.**

Enter: **Formula =25*EXP(1.386*A2) in cell B2.**

Enter: **The value 3.5 in cell A2.**

	A	B	C
1	t	N(t)	
2	0.6	=25*EXP(1.386*A2)	

	A	B
1	t	N(t)
2	3.5	3196.705

Example 3: Exponential Decay

Enter: **The value 15000 in cell A2.**

Enter: **Formula =500*EXP(-0.000124*A2) in cell B2.**

Enter: **The value 45000 in cell A2.**

	A	B	C	D
1	t	N(t)		
2	15000	=500*EXP(-0.000124*A2)		

	A	B
1	t	N(t)
2	45000	1.886283

Example 4: Depreciation

Copy the data from the text into the worksheet.

	A	B
1	x	Value ($)
2	1	12575
3	2	9455
4	3	8115
5	4	6845
6	5	5225
7	6	4485

Choose: **Chart Wizard > XY (Scatter) > Next**

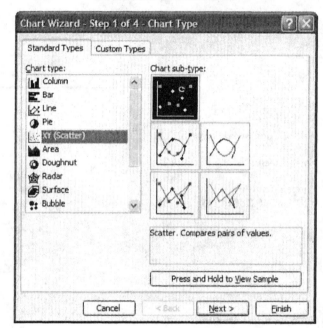

Enter: **Data Range
A3:B7 or select cells**

Select: **Series in:
columns**

Click: **Next**

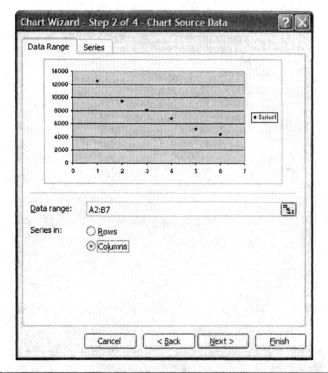

Enter: **Chart title,
Value (X) axis, and
Value (Y) axis**

Click: **Finish**

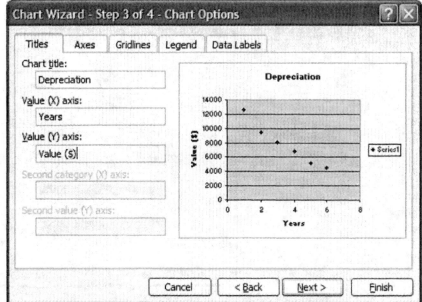

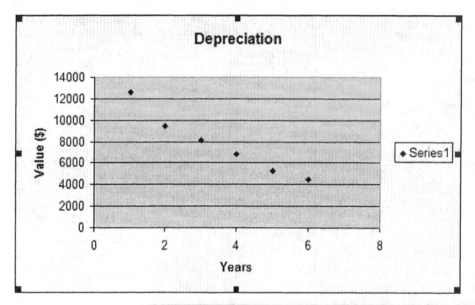

Choose: **Chart >**
Add Trendline

Select:
Exponential

Select: **Options**
Tab

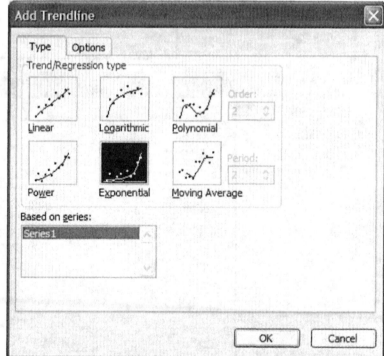

Select: **Display
equation on
chart**

Click: **OK**

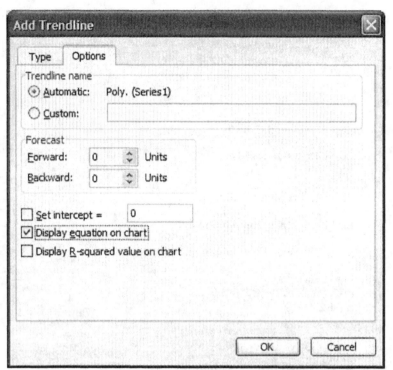

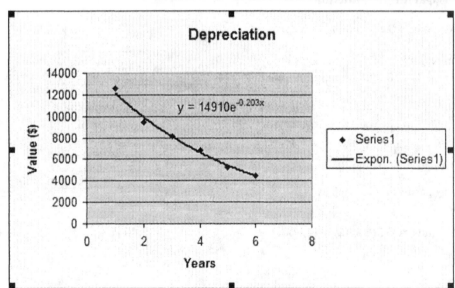

It is important to note that Excel does not generate an exponential model of the form $y = ab^x$. Instead, Excel returns a function of the form $y = ae^{bx}$. Writing this as $y = a\left(e^b\right)^x$ gives the result in the previous format $y = 14910\left(e^{-0.203}\right)^x = 14910(0.816)^x$.

Example 5: Compound Growth

Enter: **The value 1000 in cell A2.**

	A	B	C	D	E	F	G
1	P	r	m	t	A		
2	1000	0.1	12	10	=A2*(1+B2/C2)^(C2*D2)		

Enter: **The value 0.1 in cell B2.**

Enter: **The value 12 in cell C2.**

Enter: **The value 10 in cell D2.**

Enter: **Formula =A2*(1+B2/C2)^(C2*D2) in cell E2.**

	A	B	C	D	E
1	P	r	m	t	A
2	1000	0.1	12	10	2707.041

Section 2-5 Logarithmic Functions

Example 3: Solutions of the Equation $y = \log_b x$

To compute $y = \log_4 16$,
Enter: **Formula =LOG(16,4) in cell A1.**

	A
1	=LOG(16,4)

To solve $\log_2 x = -3$ for x,
Enter: **The value 2 in cell A2.**

Enter: **Any positive value (such as 1) in cell B2.**

Enter: **Formula =LOG(B2,A2) in cell C2.**

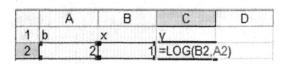

	A	B	C	D
1	b	x	y	
2	2	1	=LOG(B2,A2)	

Choose: **Tools > Goal Seek**

Enter: Set cell: **C2**

Enter: To value: **-3**

Enter: By changing cell: **B2**

Click: **OK**

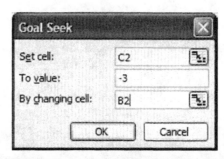

Goal Seek Status window will appear.

Click: **OK**

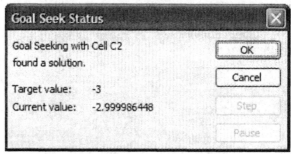

A "solution" will appear in cell **B2.** Note that this solution (0.125001) is not exact.

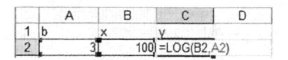

To solve $\log_b 100 = 2$ for b,

Enter: **A guess for the value of b in cell A2 (e.g. value b=3).**

Enter: **The value 100 in cell B2.**

Enter: **Formula =LOG(B2,A2) in cell C2.**

Choose: **Tools > Goal Seek**

Enter: Set cell: **C2**

Enter: To value: **2**

Enter: By changing cell: **A2**

Click: **OK**

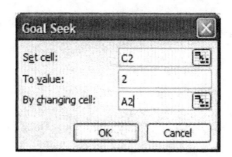

Goal Seek Status window will appear.

Click: **OK**

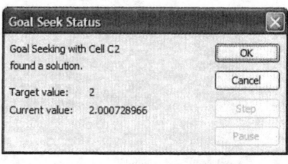

A "solution" will appear in cell **A2**. Note that this solution (9.997167) is not exact.

	A	B	C
1	b	x	y
2	9.997167	100	2.000246

Example 7: Calculator Evaluation of Logarithms

To compute $\log 3{,}184$,
Enter: **Formula =LOG(3184) in cell A1.**

	A	B
1	=LOG(3184)	

To compute $\ln 0.000349$,
Enter: **Formula =LN(0.000349) in cell A1.**

	A	B
1	=LN(0.000349)	

To compute $\log(-3.24)$,
Enter: **Formula =LOG(-3.24) in cell A1.**

	A	B
1	=LOG(-3.24)	

Since -3.24 is not in the domain of the log function, an error message occurs in cell **A1**.

	A
1	#NUM!

Example 9: Solving Exponential Equations

To solve $10^x = 2$ for x,

Enter: **A guess for the value of** x **in cell A2** (e.g. value $x=1$).

Enter: **Formula =10^A2 in cell B2.**

	A	B
1	x	10^x
2	1	=10^A2

Choose: **Tools > Goal Seek**

Enter: Set cell: **B2**

Enter: To value: **2**

Enter: By changing cell: **A2**

Click: **OK**

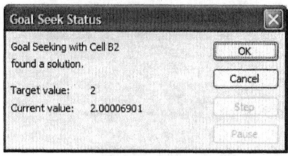

Goal Seek Status window will appear.

Click: **OK**

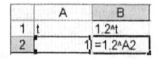

A "solution" will appear in cell A2. Note that this solution (0.301045) is not exact (the exact solution has value $\log 2$).

	A	B
1	x	10^x
2	0.301045	2.000069

Example 10: Doubling Time for an Investment

To solve $2 = 1.2^t$ for t,
Enter: **A guess for the value of *t* in cell A2** (e.g. value *t*=1).

Enter: **Formula =1.2^A2 in cell B2.**

	A	B
1	t	1.2^t
2	1	=1.2^A2

Choose: **Tools > Goal Seek**

Enter: Set cell: **B2**

Enter: To value: **2**

Enter: By changing cell: **A2**

Click: **OK**

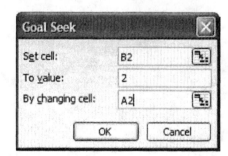

Goal Seek Status window will appear.

Click: **OK**

Goal Seek Status	
Goal Seeking with Cell B2 found a solution.	OK
	Cancel
Target value: 2	Step
Current value: 1.999700427	Pause

A "solution" will appear in cell **A2**. Note that this solution (3.800962) is not exact.

	A	B
1	t	1.2^t
2	3.800962	1.9997

Example 11: Home Ownership Rates

Copy the data from the text into the worksheet.

	A	B
1	Year	Rate (%)
2	50	55
3	60	61.9
4	70	62.9
5	80	64.4
6	90	64.2
7	100	67.4

Choose: **Chart
Wizard > XY
(Scatter) > Next**

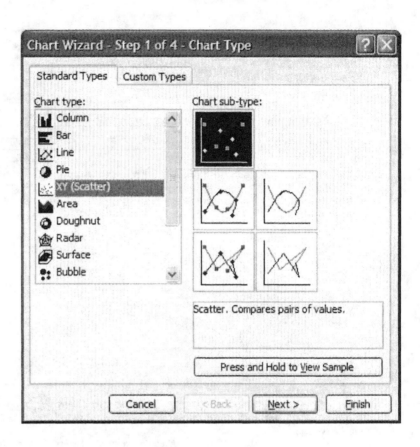

Enter: **Data Range
A2:B7 or select cells**

Select: **Series in:
columns**

Click: **Next**

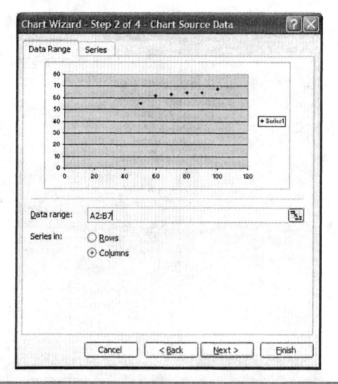

Enter: **Chart title,
Value (X) axis, and
Value (Y) axis**

Click: **Finish**

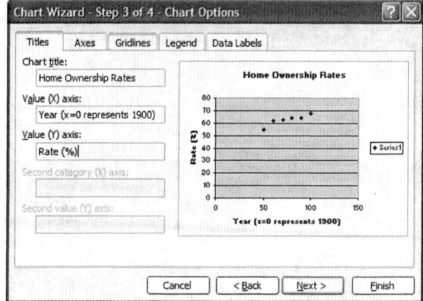

Select: **Chart**
(highlight chart)

Choose: **Chart >**
Add Trendline

Select: **Logarithmic**

Select: **Options Tab**

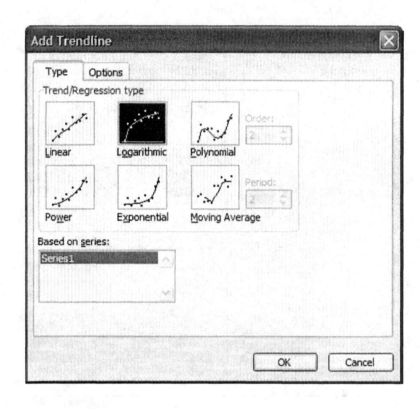

Select: **Display
equation on chart**

Click: **OK**

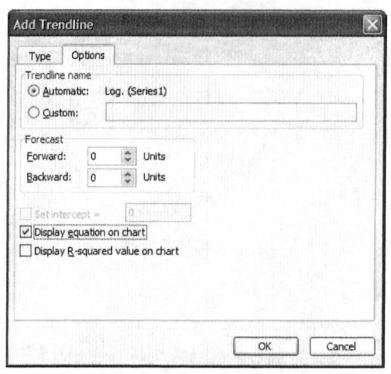

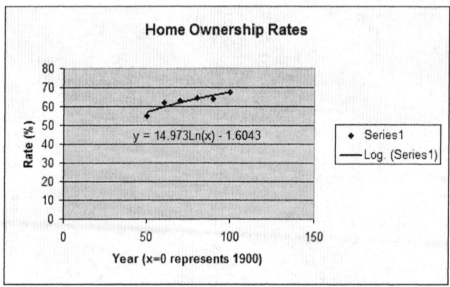

To predict the home ownership rate in 2015,

Enter: **Value 115 in cell A8**

Enter: **Formula =14.973*LN(A8)-1.6043 in cell B8**

	A	B	C	D
1	Year	Rate (%)		
2	50	55		
3	60	61.9		
4	70	62.9		
5	80	64.4		
6	90	64.2		
7	100	67.4		
8	115	=14.973*LN(A8)-1.6043		

Conclude that the home ownership rate would be 69.4% in 2015.

	A	B
1	Year	Rate (%)
2	50	55
3	60	61.9
4	70	62.9
5	80	64.4
6	90	64.2
7	100	67.4
8	115	69.44157

Chapter 3

Section 3-1 Simple Interest

Example 1: Total Amount Due on a Loan

Enter: **800** in cell **A2**

Enter: **0.09** in cell **B2**

Enter: **Formula =4/12 in cell C2**

Enter: **Formula =A2*(1+B2*C2) in cell D2**

Optional:

Select: Cell **D2**

Choose: **Format > Cells >**

Select: **Category: Currency**

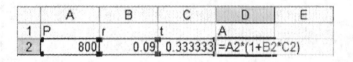

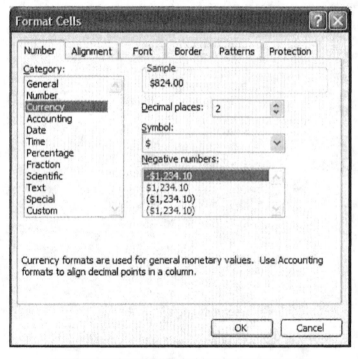

Optional:

Select: Cell **B2**

Choose: **Format > Cells >**

Select: **Category: Percentage**

	A	B	C	D
1	P	r	t	A
2	800	9.00%	0.333333	$824.00

Example 2: Present Value of an Investment

Enter: **Any value (e.g. $4000) in cell A2**

Enter: **0.1 (10%) in cell B2**

Enter: **Formula =9/12 in cell C2**

Enter: **Formula =A2*(1+B2*C2) in cell D2**

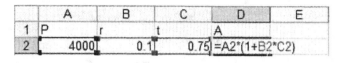

Choose: **Tools > Goal Seek**

Enter: Set cell: **D2**

Enter: To value: **5000**

Enter: By changing cell: **A2**

Click: **OK**

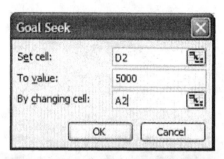

Goal Seek Status window will appear.

Click: **OK**

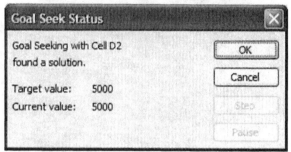

A "solution" ($4651.16) will appear in cell **A2.**

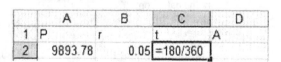

Example 3: Interest Rate Earned on a Note

Enter: **$9893.78 in cell A2**

Enter: **Any value (e.g. 0.05 or 5%) in cell B2**

Enter: **Formula =180/360 in cell C2**

Enter: **Formula =A2*(1+B2*C2) in cell D2**

Choose: **Tools > Goal Seek**

Enter: Set cell: **D2**

Enter: To value: **10000**

Enter: By changing cell: **B2**

Click: **OK**

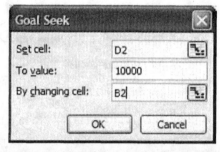

Goal Seek Status window will appear.

Click: **OK**

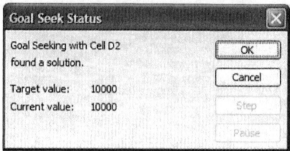

A "solution" (0.021472) will appear in cell **B2**.

	A	B	C	D
1	P	r	t	A
2	9893.78	0.021472	0.5	10000

Optional:

Select: **Cell B2**

Choose: **Format > Cells >**

Select: Category: **Percentage**

Select: Decimal Places: **3**

	A	B	C	D
1	P	r	t	A
2	9893.78	2.147%	0.5	10000

Example 5: Interest on an Investment

Copy the data from the text into the worksheet.

	A	B	C
1	Transaction Size	Fixed Commission	Variable Commission
2	$0-2,499	29	0.016
3	$2,50-$9,999	49	0.008
4	$10,000+	99	0.003

Enter: **Value 50 in cell A7**

Enter: **Value $47.52 in cell B7**

Enter: **Formula =A7*B7 in cell C7**

	A	B	C	D	E	F
1	Transaction Size	Fixed Commission	Variable Commission			
2	$0-2,499	29	0.016			
3	$2,50-$9,999	49	0.008			
4	$10,000+	99	0.003			
5						
6	Shares bought	Cost/Share	Principal	Commission	Total Investment	
7	50	47.52	=A7*B7			

Enter: **Formula =IF(C7<2500,B2+C 2*C7,IF(C7<10000, B3+C3*C7,B4+C4* C7)) in cell D7**

	A	B	C	D	E	F	G	H	I
1	Transaction Size	Fixed Commission	Variable Commission						
2	$0-2,499	29	0.016						
3	$2,50-$9,999	49	0.008						
4	$10,000+	99	0.003						
5									
6	Shares bought	Cost/Share	Principal	Commission	Total Investment				
7	50	47.52	2376	=IF(C7<2500,B2+C2*C7,IF(C7<10000,B3+C3*C7,B4+C4*C7))					

Enter: **Formula =C7+D7 in cell E7**

Optional:

Select: **Cells B7:E7**

Choose: **Format > Cells >**

Select: Category: **Currency**

	A	B	C	D	E
1	Transaction Size	Fixed Commission	Variable Commission		
2	$0-2,499	29	0.016		
3	$2,50-$9,999	49	0.008		
4	$10,000+	99	0.003		
5					
6	Shares bought	Cost/Share	Principal	Commission	Total Investment
7	50	47.52	2,376.00	67.02	2,443.02

Enter: **Value 50 in cell A10**

Enter: **Value $52.19 in cell B10**

Enter: **Formula =A10*B10 in cell C10**

Enter: **Formula =IF(C10<2500,B2+ C2*C10,IF(C10<10 000,B3+C3*C10,B4 +C4*C10)) in cell D10**

	A	B	C	D	E	F	G	H
1	Transaction Size	Fixed Commission	Variable Commission					
2	$0-2,499	29	0.016					
3	$2,50-$9,999	49	0.008					
4	$10,000+	99	0.003					
5								
6	Shares bought	Cost/Share	Principal	Commission	Total Investment			
7	50	47.52	2,376.00	67.02	2,443.02			
8								
9	Shares sold	Cost/Share	Principal	Commission	Total return			
10	50	52.19	2,609.50	=IF(C10<2500,B2+C2*C10,IF(C10<10000,B3+C3*C10,B4+C4*C10))				

Enter: **Formula =C10-D10 in cell E10**

Optional:

Select: **Cells B10:E10**

Choose: **Format > Cells >**

Select: Category: **Currency**

	A	B	C	D	E
1	Transaction Size	Fixed Commission	Variable Commission		
2	$0-2,499	29	0.016		
3	$2,50-$9,999	49	0.008		
4	$10,000+	99	0.003		
5					
6	Shares bought	Cost/Share	Principal	Commission	Total Investment
7	50	47.52	2,376.00	67.02	2,443.02
8					
9	Shares sold	Cost/Share	Principal	Commission	Total return
10	50	52.19	2,609.50	69.88	=C10-D10

Enter: **Formula =E7 in cell A13**

Enter: **Any value (e.g. 0.05) in cell B13**

Enter: **Formula =200/360 in cell C13**

	A	B	C	D	E
1	Transaction Size	Fixed Commission	Variable Commission		
2	$0-2,499	29	0.016		
3	$2,50-$9,999	49	0.008		
4	$10,000+	99	0.003		
5					
6	Shares bought	Cost/Share	Principal	Commission	Total Investment
7	50	47.52	2,376.00	67.02	2,443.02
8					
9	Shares sold	Cost/Share	Principal	Commission	Total return
10	50	52.19	2,609.50	69.88	2,539.62
11					
12	P	r	t	A	
13	2,443.02	0.05	=200/360		

Enter: **Formula =A13*(1+B13*C13) in cell D13**

	A	B	C	D	E
1	Transaction Size	Fixed Commission	Variable Commission		
2	$0-2,499	29	0.016		
3	$2,50-$9,999	49	0.008		
4	$10,000+	99	0.003		
5					
6	Shares bought	Cost/Share	Principal	Commission	Total Investment
7	50	47.52	2,376.00	67.02	2,443.02
8					
9	Shares sold	Cost/Share	Principal	Commission	Total return
10	50	52.19	2,609.50	69.88	2,539.62
11					
12	P	r	t	A	
13	2,443.02	0.05	0.555555556	=A13*(1+B13*C13)	

Choose: **Tools > Goal Seek**

Enter: Set cell: **D13**

Enter: To value: **$2539.62**

Enter: By changing cell: **B13**

Click: **OK**

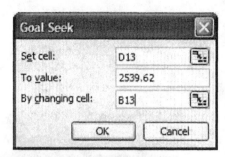

Goal Seek Status
window will appear.

Click: **OK**

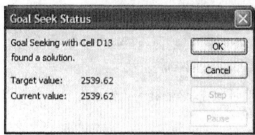

A "solution"
(7.118%) will appear
in cell **B13**.

	A	B	C	D	E
1	Transaction Size	Fixed Commission	Variable Commission		
2	$0-2,499	29	0.016		
3	$2,50-$9,999	49	0.008		
4	$10,000+	99	0.003		
5					
6	Shares bought	Cost/Share	Principal	Commission	Total Investment
7	50	47.52	2,376.00	67.02	2,443.02
8					
9	Shares sold	Cost/Share	Principal	Commission	Total return
10	50	52.19	2,609.50	69.88	2,539.62
11					
12	P	r	t	A	
13	2,443.02	7.118%	0.555555556	2539.62	

Section 3-2 Compound and Continuous Compound Interest

Example 1: Comparing Interest for Various Compounding Periods

Compounding annually:
Enter: **1000 in cell A2**

Enter: **0.08 in cell B2**

Enter: **Value 5 in cell C2**

Enter: **Formula =A2*(1+B2)^C2 in
cell D2**

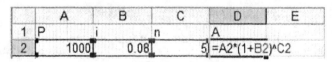

	A	B	C	D	E
1	P	i	n	A	
2	1000	0.08	5	=A2*(1+B2)^C2	

Compounding semiannually:
Enter: **1000 in cell A2**

Enter: Formula =**0.08/2 in cell B2**

Enter: **Formula =2*5 in cell C2**

Enter: **Formula =A2*(1+B2)^C2 in cell D2**

	A	B	C	D
1	P	i	n	A
2	1000	0.04	=2*5	1480.244

Compounding quarterly:
Enter: **1000 in cell A2**

Enter: Formula =**0.08/4 in cell B2**

Enter: **Formula =4*5 in cell C2**

Enter: **Formula =A2*(1+B2)^C2 in cell D2**

	A	B	C	D
1	P	i	n	A
2	1000	=0.08/4	20	1485.947

Example 2: Compounding Daily and Continuously

Enter: **The value 5000 in cell A2.**

Enter: **The value 0.08 in cell B2.**

Enter: **The value 365 in cell C2.**

Enter: **The value 2 in cell D2.**

	A	B	C	D	E	F	G
1	P	r	m	t	A		
2	5000	0.08	365	2	=A2*(1+B2/C2)^(C2*D2)		

Enter: **Formula =A2*(1+B2/C2)^(C2*D2) in cell E2.**

	A	B	C	D	E
1	P	r	m	t	A
2	5000	0.08	365	2	5867.451

Enter: **The value 5000 in cell A2.**

Enter: **The value 0.08 in cell B2.**

Enter: **The value 2 in cell C2.**

Enter: **Formula =A2*EXP(B2*C2) in cell D2.**

	A	B	C	D	E
1	P	r	t	A	
2	5000	0.08	2	=A2*EXP(B2*C2)	

	A	B	C	D
1	P	r	t	A
2	5000	0.08	2	5867.554

Example 3: Finding Present Value

Interest compounded quarterly:
Enter: **Any value (e.g. $6000) in cell A2.**

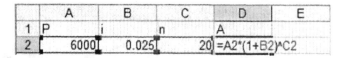

	A	B	C	D	E
1	P	i	n	A	
2	6000	0.025	20	=A2*(1+B2)^C2	

Enter: **The formula =0.10/4 (2.5%) in cell B2.**

Enter: **The formula =4*5 in cell C2.**

Enter: **The formula =A2*(1+B2)^C2 in cell D2.**

Choose: **Tools > Goal Seek**

Enter: Set cell: **D2**

Enter: To value: **$8000**

Enter: By changing cell: **A2**

Click: **OK**

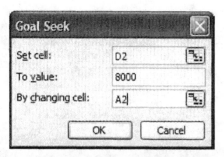

Goal Seek Status window will appear.

Click: **OK**

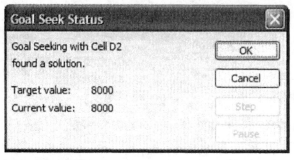

The required principal will appear in cell **A2.**

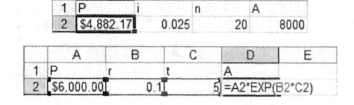

	A	B	C	D
1	P	i	n	A
2	$4,882.17	0.025	20	8000

Interest compounded continuously:
Enter: **Any value (e.g. $6000) in cell A2.**

	A	B	C	D	E
1	P	r	t	A	
2	$6,000.00	0.1	5	=A2*EXP(B2*C2)	

Enter: **The value 0.10 (10%) in**

cell **B2.**

Enter: **The value 5 in cell C2.**

Enter: **The formula =A2*EXP(B2*C2) in cell D2.**

Choose: **Tools > Goal Seek**

Enter: Set cell: **D2**

Enter: To value: **$8000**

Enter: By changing cell: **A2**

Click: **OK**

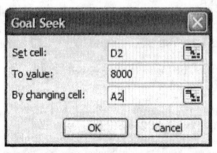

Goal Seek Status window will appear.

Click: **OK**

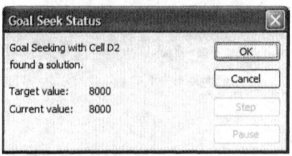

The required principal will appear in cell **A2.**

	A	B	C	D
1	P	r	t	A
2	$4,852.25	0.1	5	8000

Example 4: Computing Growth Rate

Interest compounded annually:
Enter: **The value 10000 in cell A2.**

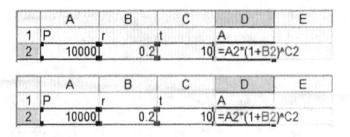

	A	B	C	D	E
1	P	r	t	A	
2	10000	0.2	10	=A2*(1+B2)^C2	

	A	B	C	D	E
1	P	r	t	A	
2	10000	0.2	10	=A2*(1+B2)^C2	

Enter: **Any value (e.g. 0.2) in cell B2.**

Enter: **The value 10 in cell C2.**

Enter: **The formula**

=A2*(1+B2)^C2 in cell D2.

Choose: **Tools > Goal Seek**

Enter: Set cell: **D2**

Enter: To value: **$126000**

Enter: By changing cell: **B2**

Click: **OK**

Goal Seek Status window will appear.

Click: **OK**

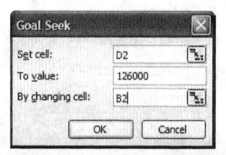

The growth rate (28.836%) will appear in cell **B2.**

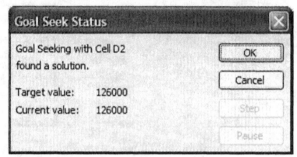

Interest compounded continuously:
Enter: **The value 10000 in cell A2.**

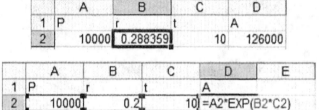

Enter: **Any value (e.g. 0.2) in cell B2.**

Enter: **The value 10 in cell C2.**

Enter: **The formula =A2*EXP(B2*C2) in cell D2.**

Choose: **Tools > Goal Seek**

Enter: Set cell: **D2**

Enter: To value: **$126000**

Enter: By changing cell: **B2**

Click: **OK**

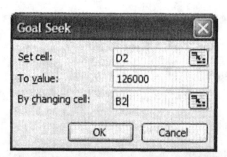

Goal Seek Status window will appear.

Click: **OK**

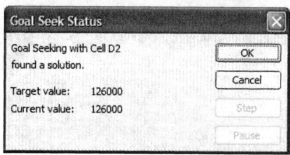

The growth rate (25.337%) will appear in cell **B2**.

	A	B	C	D
1	P	r	t	A
2	10000	0.25337	10	126000

Example 5: Computing Growth Time

Enter: **The value 10000 in cell A2.**

Enter: **The formula =0.09/12 in cell B2.**

Enter: **Any value (e.g. 12 months) in cell C2.**

Enter: **The formula =A2*(1+B2)^C2 in cell D2.**

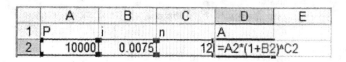

	A	B	C	D	E
1	P	i	n	A	
2	10000	0.0075	12	=A2*(1+B2)^C2	

Choose: **Tools > Goal Seek**

Enter: Set cell: **D2**

Enter: To value: **$12000**

Enter: By changing cell: **C2**

Click: **OK**

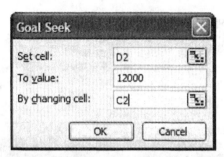

Goal Seek Status window will appear.

Click: **OK**

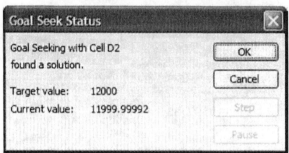

The growth time will appear in cell **C2.** Since the growth time must be a positive integer in this example, 24.4 is rounded up to 25 months.

	A	B	C	D
1	P	i	n	A
2	10000	0.0075	24.40059	12000

Example 6: Using APY to Compare Investments

Copy the data from the text into the worksheet.

	A	B	C
1	Bank	Rate	m
2	Advanta	4.93%	12
3	DeepGreen	4.95%	365
4	Charter One	4.97%	4
5	Liberty	4.94%	

Enter: **Formula =(1+B2/C2)^C2-1 in cell D2**

	A	B	C	D	E
1	Bank	Rate	m	APY	
2	Advanta	4.93%	12	=(1+B2/C2)^C2-1	
3	DeepGreen	4.95%	365		
4	Charter One	4.97%	4		
5	Liberty	4.94%			

Copy: **Contents of cell D2 to cells D3:D4**

	A	B	C	D
1	Bank	Rate	m	APY
2	Advanta	4.93%	12	0.050429
3	DeepGreen	4.95%	365	0.050742
4	Charter One	4.97%	4	0.050634
5	Liberty	4.94%		

Enter: **Formula =EXP(B2)-1 in cell D5**

	A	B	C	D
1	Bank	Rate	m	APY
2	Advanta	4.93%	12	0.050429372
3	DeepGreen	4.95%	365	0.050742066
4	Charter One	4.97%	4	0.05063398
5	Liberty	4.94%		=EXP(B2)-1

Optional:

Select: **Cells D2:D5**

Choose: **Format > Cells >**

Select: Category: **Percentage**

Select: Decimal Places: **3**

	A	B	C	D
1	Bank	Rate	m	APY
2	Advanta	4.93%	12	5.043%
3	DeepGreen	4.95%	365	5.074%
4	Charter One	4.97%	4	5.063%
5	Liberty	4.94%		5.054%

Example 7: Computing the Annual Nominal Rate Given the APY

Enter: **Any value (0.05) in cell A2**

Enter: **The value 12 in cell B2**

Enter: **The formula =(1+A2/B2)^B2-1 in cell C2**

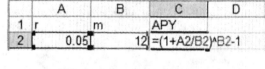

	A	B	C	D
1	r	m	APY	
2	0.05	12	=(1+A2/B2)^B2-1	

Choose: **Tools > Goal Seek**

Enter: Set cell: **C2**

Enter: To value: **0.075**

Enter: By changing cell: **A2**

Click: **OK**

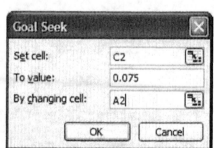

Goal Seek Status window will appear.

Click: **OK**

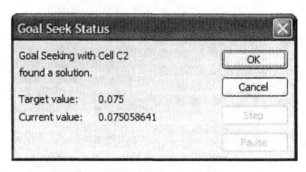

The (approximate) annual nominal rate will appear in cell **A2**. Note that this actually produces an APY of 7.5059%.

	A	B	C
1	r	m	APY
2	0.072594	12	0.075059

Section 3-3 Future Value of an Annuity; Sinking Funds

Example 1: Future Value of an Ordinary Annuity

Enter: **The value 0.085 in cell A2.**

Enter: **The value 20 in cell B2.**

Enter: **The value 2000 in cell C2.**

	A	B	C	D	E
1	i	n	PMT	FV	
2	0.085	20	2000	=FV(A2,B2,C2)	

Enter: **The formula =FV(A2,B2,C2) in cell D2.** [FV is Excel's built-in Future Value function]

Note: Output is negative. Making the PMT negative will make FV positive. It's all a matter of perspective.

	A	B	C	D
1	i	n	PMT	FV
2	0.085	20	2000	($96,754.03)

Example 2: Computing the Payment for a Sinking Fund

Enter: **The formula =0.066/12 in cell A2**

Enter: **The value 60 in cell B2**

	A	B	C	D	E
1	i	n	PMT	FV	
2	0.0055	60	10000	=FV(A2,B2,C2)	

Enter: **Any value (e.g. $10000) in cell C2**

Enter: **Formula =FV(A2,B2,C2) in cell D2.**

Choose: **Tools > Goal Seek**

Enter: Set cell: **D2**

Enter: To value: **800000**

Enter: By changing cell: **C2**

Click: **OK**

Goal Seek Status window will appear.

Click: **OK**

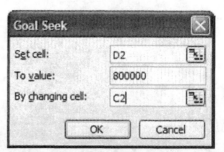

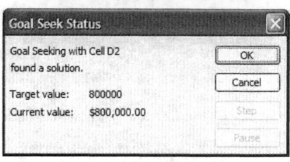

The monthly payment ($11,290.42) will appear in cell **C2**. Note that this number is negative (this amount is paid into an account each month).

	A	B	C	D
1	i	n	PMT	FV
2	0.0055	60	-11290.42	$800,000.00

Example 4: Approximating an Interest Rate

Enter: **Any value (e.g. 0.005) in cell A2.**
Note: This value should be reasonably close to the actual monthly interest rate.

Enter: **The value 360 in cell B2.**

Enter: **The value -100 in cell C2.**

Enter: **The formula =FV(A2,B2,C2) in cell D2.**

	A	B	C	D
1	i	n	PMT	FV
2	0.005	360	-100	=FV(A2,B2,C2)

Choose: **Tools > Goal Seek**

Enter: Set cell: **D2**

Enter: To value: **160000**

Enter: By changing cell: **A2**

Click: **OK**

Goal Seek Status window will appear.

Click: **OK**

The monthly interest rate (0.6957%) will appear in cell **A2**. To convert this to an annual rate, multiply this number by 12 (to get 8.35%).

	A	B	C	D
1	i	n	PMT	FV
2	0.006957	360	-100	$160,000.00

Section 3-4 Present Value of an Annuity; Amortization

Example 1: Present Value of an Annuity

Enter: **The formula =0.06/12 in cell A2.**

Enter: **The formula =5*12 in cell B2.**

Enter: **The value 200 in cell C2.**

Enter: **The formula =PV(A2,B2,C2) in cell D2.** [PV is Excel's built-in Present Value function]

	A	B	C	D	E
1	i	n	PMT	PV	
2	0.005	60	200	=PV(A2,B2,C2)	

Note: Output is negative. Making the PMT negative will make PV positive. It's all a matter of perspective.

	A	B	C	D
1	i	n	PMT	PV
2	0.005	60	200	($10,345.11)

Example 2: Retirement Planning

Enter: **The value 0.065 in cell A2.**

Enter: **The value 20 in cell B2.**

Enter: **The value -25000 in cell C2.**

	A	B	C	D	E
1	i	n (withdrawals)	PMT (out)	PV	
2	0.065	20	-25000	=PV(A2,B2,C2)	

Enter: **The formula =PV(A2,B2,C2) in cell D2.**

Enter: **The formula =A2 in cell A5.**

Enter: **The value 25 in cell B5.**

Enter: **The formula =D2 in cell C5.**

	A	B	C	D	E
1	i	n (withdrawals)	PMT (out)	PV	
2	0.065	20	-25000	$275,462.68	
3					
4	i	n (deposits)	FV	PMT (in)	
5	0.065	25	$275,462.68	=PMT(A5,B5,0,C5)	

Enter: **The formula =PMT(A5,B5,0,C5) in cell D5.** PMT is Excel's built-in payment function.

Example 3: Monthly Payment and Total Interest on an Amortized Debt

Enter: **The value 0.015 (1.5%) in cell A2.**

Enter: **The value 18 in cell B2.**

Enter: **The value 800 in cell C2.**

	A	B	C	D	E
1	i	n	PV	PMT	
2	0.015	18	800	=PMT(A2,B2,C2)	

Enter: **The formula =PMT(A2,B2,C2) in cell D2.**

Note: The value of PMT is negative.

Enter: **The formula =B2*(-D2)-C2 in cell E5.**

	A	B	C	D	E
1	i	n	PV	PMT	Total Interest
2	0.015	18	800	($51.04)	=B2*(-D2)-C2

Example 4: Constructing an Amortization Schedule

Enter: **The value 0.01 (1%) in cell A2.**

Enter: **The value 6 in cell B2.**

	A	B	C	D	E
1	i	n	PV	PMT	
2	0.01	6	-500	=PMT(A2,B2,C2)	

Enter: **The value -500 (amount**

borrowed) in cell C2.

Enter: **The formula =PMT(A2,B2,C2) in cell D2.**

	A	B	C	D	E
1	i	n	PV	PMT	
2	0.01	6	-500	=PMT(A2,B2,C2)	

Enter: **The value 0 in cell A5**

Enter: **The formula =-C2 in cell E5**

	A	B	C	D	E
1	i	n	PV	PMT	
2		0.01	6	-500	$86.27
3					
4	Payment Number	Payment	Interest	Unpaid Balance Reduction	Unpaid Balance
5	0				=-C2

Enter: **The formula =A5+1 in cell A6**

Enter: **The formula =$D$2 (an absolute reference) in cell B6**

	A	B	C	D	E
1	i	n	PV	PMT	
2		0.01	6	-500	$86.27
3					
4	Payment Number	Payment	Interest	Unpaid Balance Reduction	Unpaid Balance
5	0				500
6	1	=$D$2			

Enter: **The formula =$A$2*E5 in cell C6**

	A	B	C	D	E
1	i	n	PV	PMT	
2	0.01	6	-500	$86.27	
3					
4	Payment Number	Payment	Interest	Unpaid Balance Reduction	Unpaid Balance
5	0				500
6	1	$86.27	=$A$2*E5		

Enter: **The formula =B6-C6 in cell D6**

Enter: **The formula =E5-D6 in cell E6**

	A	B	C	D	E
1	i	n	PV	PMT	
2		0.01	6	-500	$86.27
3					
4	Payment Number	Payment	Interest	Unpaid Balance Reduction	Unpaid Balance
5	0				$500.00
6	1	$86.27	$5.00	$81.27	=E5-D6

Select: **Cells A6:E6**

Copy: **Values of cells A6:E6 to cells A7:E11** (drag lower right hand corner down using mouse)

	A	B	C	D	E
1	i	n	PV	PMT	
2		0.01	6	-500	$86.27
3					
4	Payment Number	Payment	Interest	Unpaid Balance Reduction	Unpaid Balance
5	0				$500.00
6	1	$86.27	$5.00	$81.27	$418.73
7	2	$86.27	$4.19	$82.09	$336.64
8	3	$86.27	$3.37	$82.91	$253.73
9	4	$86.27	$2.54	$83.74	$169.99
10	5	$86.27	$1.70	$84.57	$85.42
11	6	$86.27	$0.85	$85.42	$0.00

Chapter 4

Section 4-1 Review: Systems of Linear Equations in Two Variables

Example 3: Solving a System using Excel

Enter: **Any value (e.g. 0) in cell A2**

Enter: **Any value (e.g. 0) in cell B2**

Enter: **Formula =5*A2+2*B2 in cell A5** (representing the left hand side of equation 1)

Enter: **The value 15 in cell B5** (representing the right hand side of equation 1)

Enter: **Formula =2*A2-3*B2 in cell A6** (representing the left hand side of equation 2)

Enter: **The value 16 in cell B6** (representing the right hand side of equation 2)

Choose: **Tools > Solver >**

Note: If Solver is not an option, it may need to be installed. If so,

Choose: **Tools > Add-ins >**

Select: **Solver Add-in**

Click: **OK**

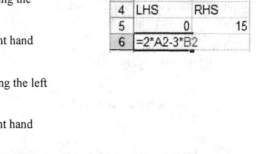

	A	B
1	x	y
2	0	0
3		
4	LHS	RHS
5	0	15
6	=2*A2-3*B2	

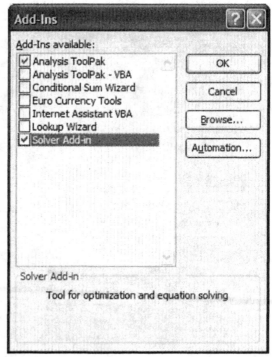

If this add-in has not been installed before, you will be prompted to install it now.

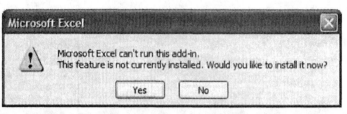

Choose: **Tools** > **Solver** >

Enter: Select Target Cell: **A2**

Select: Equal To: **Max (or Min)**

Enter: By Changing Cells: **A2, B2**

Click: Subject to the Constraints: **Add**

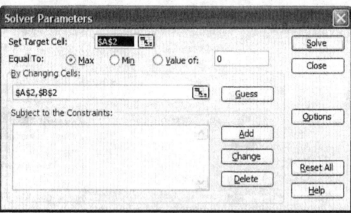

Enter: Cell Reference: **A5**

Select: =

Enter: Constraint: **B5**

Click: **Add**

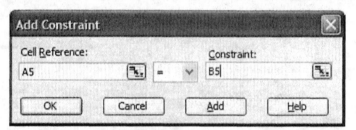

Enter: Cell Reference: **A6**

Select: =

Enter: Constraint: **B6**

Click: **OK**

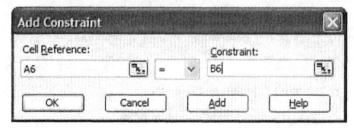

Click: **Solve**

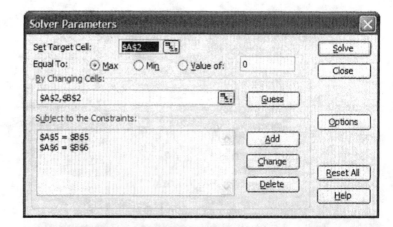

Select: **Keep Solver Solution**

Click: **OK**

Since exactly one solution exists, an approximation to it appears in cells **A2** and **B2**.

	A	B
1	x	y
2	4.052632	-2.63158
3		
4	LHS	RHS
5	15	15
6	16	16

Example 7: Supply and Demand

Enter: **Any value (e.g. 0) in cell A2**

Enter: **Any value (e.g. 0) in cell B2**

Enter: **The formula =A2 in cell A5** (representing the left hand side of equation 1)

Enter: **The formula =-0.2*B2+4 in cell B5** (representing the right hand side of equation 1)

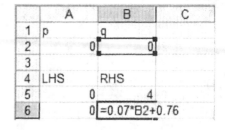

	A	B	C
1	p	q	
2		0	0
3			
4	LHS	RHS	
5	0	4	
6	0	=0.07*B2+0.76	

Enter: **The formula =A2 in cell A6** (representing the left hand side of equation 2)

Enter: **The formula =0.07*B2+0.76 in cell B6** (representing the right hand side of equation 2)

Choose: **Tools > Solver >**

Enter: Select Target Cell: **A2**

Select: Equal To: **Max**

Enter: By Changing Cells: **A2, B2**

Click: Subject to the Constraints: **Add**

* If Solver does not appear, see Example 3 from Section 4-1.

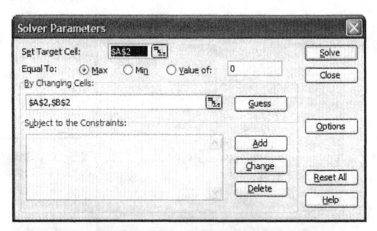

Enter: Cell Reference: **A5**

Select: =

Enter: Constraint: **B5**

Click: **Add**

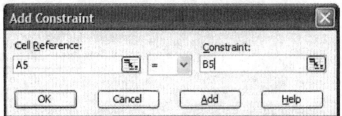

Enter: Cell Reference: **A6**

Select: =

Enter: Constraint: **B6**

Click: **OK**

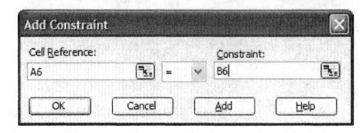

Click: **Solve**

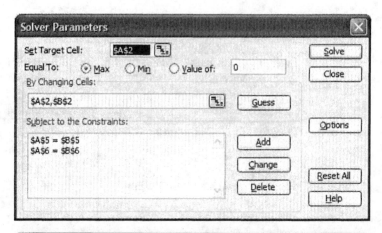

Select: **Keep Solver Solution**

Click: **OK**

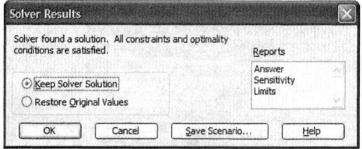

Since exactly one solution exists, an
approximation to it appears in cells
A2 and **B2**.

	A	B
1	p	q
2	1.6	12
3		
4	LHS	RHS
5	1.6	1.6
6	1.6	1.6

Section 4-2 Systems of Linear Equations and Augmented Matrices

Example 1: Solving a System Using Augmented Matrix Methods

[Perform row operation $R_1 \leftrightarrow R_2$]

Enter: **Contents of augmented matrix in cells A1:C2**

Enter: **Formula =A2 in cell A4** (this is the first step in switching the two rows)

Copy: **Contents of cell A4 to cells B4:C4**

	A	B	C
1	3	4	1
2	1	-2	7
3			
4	=A2		

Enter: **Formula =A1 in cell A5**

Copy: **Contents of cell A5 to cells B5:C5** (this completes the switching of the two rows)

	A	B	C
1	3	4	1
2	1	-2	7
3			
4	1	-2	7
5	3	4	1

[Perform row operation $(-3)R_1 + R_2 \rightarrow R_2$]
Enter: **Formula =A4 in cell A7**

Copy: **Contents of cell A7 to cells B7:C7**

Enter: **Formula =(-3)*A4+A5 in cell A8**

Copy: **Contents of cell A8 to cells B8:C8**

	A	B	C
1	3	4	1
2	1	-2	7
3			
4	1	-2	7
5	3	4	1
6			
7	1	-2	7
8	=(-3)*A4+A5		

[Perform row operation $\frac{1}{10}R_2 \rightarrow R_2$]
Enter: **Formula =A7 in cell A10**

Copy: **Contents of cell A10 to cells B10:C10**

Enter: **Formula =1/10*A8 in cell A11**

Copy: **Contents of cell A11 to cells B11:C11**

	A	B	C
1	3	4	1
2	1	-2	7
3			
4	1	-2	7
5	3	4	1
6			
7	1	-2	7
8	0	10	-20
9			
10	1	-2	7
11	=1/10*A8		

[Perform row operation $2R_2 + R_1 \rightarrow R_1$]
Enter: **Formula =2*A11+A10 in cell A13**

Copy: **Contents of cell A13 to cells B13:C13**

Enter: **Formula =A11 in cell A14**

Copy: **Contents of cell A14 to cells B14:C14**

	A	B	C
1	3	4	1
2	1	-2	7
3			
4	1	-2	7
5	3	4	1
6			
7	1	-2	7
8	0	10	-20
9			
10	1	-2	7
11	0	1	-2
12			
13	=2*A11+A10		

The reduced augmented matrix can be found in cells
A13:C14.

	A	B	C
1	3	4	1
2	1	-2	7
3			
4	1	-2	7
5	3	4	1
6			
7	1	-2	7
8	0	10	-20
9			
10	1	-2	7
11	0	1	-2
12			
13	1	0	3
14	0	1	-2

Example 3: Solving a System Using Augmented Matrix Methods

[Perform row operations $\frac{1}{2}R_1 \rightarrow R_1$ and $\frac{1}{3}R_2 \rightarrow R_2$]
Enter: **Contents of augmented matrix in cells A1:C2**

Enter: **Formula =1/2*A1 in cell A4**

Copy: **Contents of cell A4 to cells B4:C4**

Enter: **Formula =1/3*A2 in cell A5**

	A	B	C
1	2	-1	4
2	-6	3	-12
3			
4	1	-0.5	2
5	=1/3*A2		

Copy: **Contents of cell A5 to cells B5:C5**

[Perform row operation $2R_1 + R_2 \rightarrow R_2$]
Enter: **Formula =A4 in cell A7**

Copy: **Contents of cell A7 to cells B7:C7**

Enter: **Formula =2*A4+A5 in cell A8**

Copy: **Contents of cell A8 to cells B8:C8**

	A	B	C
1	2	-1	4
2	-6	3	-12
3			
4	1	-0.5	2
5	-2	1	-4
6			
7	1	-0.5	2
8	=2*A4+A5		

Even using Excel, one must recognize that there are an infinite number of solutions to this system and use the reduced augmented matrix to describe these.

	A	B	C
1	2	-1	4
2	-6	3	-12
3			
4	1	-0.5	2
5	-2	1	-4
6			
7	1	-0.5	2
8	0	0	0

Section 4-3 Gauss-Jordan Elimination

Example 3: Solving a System Using Gauss-Jordan Elimination

[Perform row operation $\frac{1}{2}R_1 \rightarrow R_1$]

Enter: **Contents of augmented matrix in cells A1:D3**

Enter: **Formula =1/2*A1 in cell A5**

Copy: **Contents of cell A5 to cells B5:D5**

Enter: **Formula =A2 in cell A6 and formula =A3 in cell A7**

Copy: **Contents of cell A6 to cells B6:D6 and contents of cell A7 to cells B7:D7**

	A	B	C	D
1	2	-4	1	-4
2	4	-8	7	2
3	-2	4	-3	5
4				
5	=1/2*A1			

[Perform row operations $(-4)R_1 + R_2 \rightarrow R_2$ and $2R_1 + R_3 \rightarrow R_3$]

Enter: **Formula =A5 in cell A9**

Copy: **Contents of cell A9 to cells B9:D9**

Enter: **Formula =(-4)*A5+A6 in cell A10**

Copy: **Contents of cell A10 to cells B10:D10**

Enter: **Formula =2*A5+A7 in cell A11**

Copy: **Contents of cell A11 to cells B11:D11**

[Perform row operation $0.2R_2 \rightarrow R_2$]

Enter: **Formula =A9 in cell A13**

Copy: **Contents of cell A13 to cells B13:D13**

Enter: **Formula =0.2*A10 in cell A14**

Copy: **Contents of cell A14 to cells B14:D14**

Enter: **Formula =A11 in cell A15**

Copy: **Contents of cell A15 to cells B15:D15**

	A	B	C	D
1	2	-4	1	-4
2	4	-8	7	2
3	-2	4	-3	5
4				
5	1	-2	0.5	-2
6	4	-8	7	2
7	-2	4	-3	5
8				
9	1	-2	0.5	-2
10	=-4*A5+A6			

	A	B	C	D
1	2	-4	1	-4
2	4	-8	7	2
3	-2	4	-3	5
4				
5	1	-2	0.5	-2
6	4	-8	7	2
7	-2	4	-3	5
8				
9	1	-2	0.5	-2
10	0	0	5	10
11	0	0	-2	1
12				
13	1	-2	0.5	-2
14	=0.2*A10			
15	0	0	-2	1

[Perform row operations $0.5R_2 + R_1 \rightarrow R_1$ and $2R_2 + R_3 \rightarrow R_3$]

Enter: **Formula =0.5*A14+A13 in cell A17**

Copy: **Contents of cell A17 to cells B17:D17**

Enter: **Formula =A14 in cell A18**

Copy: **Contents of cell A18 to cells B18:D18**

Enter: **Formula =2*A14+A15 in cell A19**

Copy: **Contents of cell A19 to cells B19:D19**

	A	B	C	D
1	2	-4	1	-4
2	4	-8	7	2
3	-2	4	-3	5
4				
5	1	-2	0.5	-2
6	4	-8	7	2
7	-2	4	-3	5
8				
9	1	-2	0.5	-2
10	0	0	5	10
11	0	0	-2	1
12				
13	1	-2	0.5	-2
14	0	0	1	2
15	0	0	-2	1
16				
17	1	-2	0	-3
18	0	0	1	2
19	=2*A14+A15			

Since the last row produces a contradiction, Gauss-Jordan elimination is stopped.

Section 4-4 Matrices: Basic Operations

Example 1: Matrix Addition

Enter: **Contents of each matrix in cells A1:C3 and cells A4:C5**

Enter: **Formula =A1+A4 in cell A7**

	A	B	C
1	2	-3	0
2	1	2	-5
3			
4	3	1	2
5	-3	2	5
6			
7	=A1+A4		

Copy: **Contents of cell A7 to cells A7:C8**

	A	B	C
1	2	-3	0
2	1	2	-5
3			
4	3	1	2
5	-3	2	5
6			
7	5	-2	2
8	-2	4	0

Example 2: Matrix Subtraction

Enter: **Contents of each matrix in cells A1:B2 and cells A4:B5**

Enter: **Formula =A1-A4 in cell A7**

	A	B
1	3	-2
2	5	0
3		
4	-2	2
5	3	4
6		
7	=A1-A4	

Copy: **Contents of cell A7 to cells A7:B8**

	A	B
1	3	-2
2	5	0
3		
4	-2	2
5	3	4
6		
7	5	-4
8	2	-4

Example 4: Multiplication of a Matrix by a Number

Enter: **Contents of the matrix in cells A1:C3**

Enter: **Formula =2*A1 in cell A5**

	A	B	C
1	3	-1	0
2	-2	1	3
3	0	-1	-2
4			
5	=-2*A1		

Copy: **Contents of cell A5 to cells A5:C7**

	A	B	C
1	3	-1	0
2	-2	1	3
3	0	-1	-2
4			
5	-6	2	0
6	4	-2	-6
7	0	2	4

Example 5: Sales Commissions

Enter: **Contents of the matrices in cells B3:C4 and D3:E4**

Enter: **Formula =B3+D3 in cell B6**

	A	B	C	D	E
1		August Sales		September Sales	
2		Compact	Luxury	Compact	Luxury
3	Smith	$54,000	$88,000	$228,000	$368,000
4	Jones	$126,000	$0	$304,000	$322,000
5		Combined Sales			
6	Smith	=B3+D3			
7	Jones				

Copy: **Contents of cell B6 to cells B6:C7**

	A	B	C	D	E
1		August Sales		September Sales	
2		Compact	Luxury	Compact	Luxury
3	Smith	$54,000	$88,000	$228,000	$368,000
4	Jones	$126,000	$0	$304,000	$322,000
5		Combined Sales			
6	Smith	$282,000	$456,000		
7	Jones	$430,000	$322,000		

Enter: **Formula =D3-B3 in cell D6**

Copy: **Contents of cell D6 to cells D6:E7**

	A	B	C	D	E
1		August Sales		September Sales	
2		Compact	Luxury	Compact	Luxury
3	Smith	$54,000	$88,000	$228,000	$368,000
4	Jones	$126,000	$0	$304,000	$322,000
5		Combined Sales		Sales Increase	
6	Smith	$282,000	$456,000	$174,000	$280,000
7	Jones	$430,000	$322,000	$178,000	$322,000

Enter: **Formula =0.05*D3 in cell F3**

Copy: **Contents of cell F3 to cells F3:G4**

	A	B	C	D	E	F	G
1		August Sales		September Sales		September Commissions	
2		Compact	Luxury	Compact	Luxury	Compact	Luxury
3	Smith	$54,000	$88,000	$228,000	$368,000	=0.05*D3	
4	Jones	$126,000	$0	$304,000	$322,000		
5		Combined Sales		Sales Increase			
6	Smith	$282,000	$456,000	$174,000	$280,000		
7	Jones	$430,000	$322,000	$178,000	$322,000		

Example 8: Matrix Multiplication

Enter: **Contents of the matrices in cells A1:B3 and A5:D6**

Select: **Cells A8:D10**

Enter: **The array formula =MMULT(A1:B3,A5:D6) in selected region** (not in just a single cell).

*Use CRTL-SHIFT-ENTER rather than ENTER to input an array formula.

	A	B	C	D
1	2	1		
2	1	0		
3	-1	2		
4				
5	1	-1	0	1
6	2	1	2	0
7				
8	=MMULT(A1:B3,A5:D6)			
9				
10				

The resulting matrix will appear in cells **A8:D10**.

	A	B	C	D
1	2	1		
2	1	0		
3	-1	2		
4				
5	1	-1	0	1
6	2	1	2	0
7				
8	4	-1	2	2
9	1	-1	0	1
10	3	3	4	-1

Example 9: Labor Costs

Enter: **Contents of the matrices in cells B3:C4 and E3:F4**

	A	B	C	D	E	F
1		Labor-hours per ski			Hourly wages	
2		Assembly	Finishing		California	Wisconsin
3	Trick ski	5	1.5	Assembly	$12	$13
4	Slalom ski	3	1	Finishing	$7	$8

Select: **Cells B7:C8**

Enter: **The array formula =MMULT(B3:C4,E3:F4) in** selected region (not in just a single cell).

*Use CRTL-SHIFT-ENTER rather than ENTER to input an array formula.

	A	B	C	D	E	F
1		Labor-hours per ski			Hourly wages	
2		Assembly	Finishing		California	Wisconsin
3	Trick ski	5	1.5	Assembly	$12	$13
4	Slalom ski	3	1	Finishing	$7	$8
5		Labor costs per ski				
6		California	Wisconsin			
7	Trick ski	=MMULT(B3:C4,E3:F4)				
8	Slalom ski					

Section 4-5 Inverse of a Square Matrix

Example 2: Finding the Inverse of a Matrix

Enter: **Contents of the matrix in cells A1:C3**

Select: **Cells A5:C7**

Enter: **The array formula =MINVERSE(A1:C3) in selected region.**

*Use CRTL-SHIFT-ENTER rather than ENTER to input an array formula.

	A	B	C
1	1	-1	1
2	0	2	-1
3	2	3	0
4			
5	=MINVERSE(A1:C3)		
6			
7			

	A	B	C
1	1	-1	1
2	0	2	-1
3	2	3	0
4			
5	3	3	-1
6	-2	-2	1
7	-4	-5	2

99

Example 4: Finding a Matrix Inverse

Enter: **Contents of the matrix in cells A1:B2**

Select: **Cells A4:B5**

Enter: **The array formula =MINVERSE(A1:B2) in selected region.**

	A	B
1	2	-4
2	-3	6
3		
4	=MINVERSE(A1:B2)	
5		

*Use CRTL-SHIFT-ENTER rather than ENTER to input an array formula.

Since this matrix is singular (does not have an inverse), Excel outputs an error message.

	A	B
1	2	-4
2	-3	6
3		
4	#NUM!	#NUM!
5	#NUM!	#NUM!

Section 4-6 Matrix Equations and Systems of Linear Equations

Example 2: Using Inverses to Solve Systems of Equations

Enter: **Contents of the coefficient matrix A in cells A1:C3**

Enter: **Contents of the matrix B in cells E1:E3**

Select: **Cells A5:A7**

Enter: **The array formula =MMULT(MINVERSE(A1:C3),E1:E3) in selected region.**

	A	B	C	D	E
1	1	-1	1		1
2	0	2	-1		1
3	2	3	0		1
4					
5	=MMULT(MINVERSE(A1:C3),E1:E3)				
6					
7					

*Use CRTL-SHIFT-ENTER rather than ENTER to input an array formula.

Note that it is important to select the right sized region to place the array formula.

	A	B	C	D	E
1	1	-1	1		1
2	0	2	-1		1
3	2	3	0		1
4					
5	5				
6	-3				
7	-7				

Example 3: Using Inverses to Solve Systems of Equations

Enter: **Contents of the coefficient matrix** A **in cells A1:C3**

Enter: **Contents of the matrix** B **in cells E1:E3**

Select: **Cells A5:A7**

Enter: **The array formula =MMULT(MINVERSE(A1:C3),E1:E3) in selected region.**

	A	B	C	D	E
1	1	-1	1		1
2	0	2	-1		1
3	2	3	0		1
4					
5	=MMULT(MINVERSE(A1:C3),E1:E3)				
6					
7					

Use **CRTL-SHIFT-ENTER to input an array formula.**

	A	B	C	D	E
1	1	-1	1		1
2	0	2	-1		1
3	2	3	0		1
4					
5	5				
6	-3				
7	-7				

To solve the second system (part (B)), simply change the values in cells **E1:E3**. The solution found in cells **A5:A7** will automatically update.

	A	B	C	D	E
1	1	-1	1		-5
2	0	2	-1		2
3	2	3	0		-3
4					
5	-6				
6	3				
7	4				

Example 4: Investment Analysis

Enter: Data for the problem

	A	B	C	D	E	F	G
1			Clients				
2		1	2	3			
3	Total Investment	$20,000	$50,000	$10,000		1	1
4	Annual Return	$2,400	$7,500	$1,300		0.1	0.2
5	Amount Invested in A						
6	Amount Invested in B						

Select: **Cells B5:B6**

Enter: **The array formula =MMULT(MINVERSE($F$3 :$G$4),B3:B4) in selected region. Use CRTL-SHIFT-ENTER to input an array formula.**

	A	B	C	D	E	F	G
1			Clients				
2		1	2	3			
3	Total Investment	$20,000	$50,000	$10,000		1	1
4	Annual Return	$2,400	$7,500	$1,300		0.1	0.2
5	Amount Invested in A	=MMULT(MINVERSE($F$3:$G$4),B3:B4)					
6	Amount Invested in B						

Copy: **Contents of cells B5:B6 to cells C5:D6**

(Select cells **B5:B6** and drag bottom right corner two cells to right)

	A	B	C	D	E	F	G
1			Clients				
2		1	2	3			
3	Total Investment	$20,000	$50,000	$10,000		1	1
4	Annual Return	$2,400	$7,500	$1,300		0.1	0.2
5	Amount Invested in A	16000	25000	7000			
6	Amount Invested in B	4000	25000	3000			

Section 4-7 Leontief Input-Output Analysis

Example 1: Input-Output Analysis

Enter: The technology matrix *M* and the final demand data for the problem

	A	B	C	D	E	F
1		Technology Matrix M				Final Demand
2		A	E	M		
3	A	0.2	0	0.1		33
4	E	0.4	0.2	0.1		37
5	M	0	0.4	0.3		64

Enter: **Matrix *I-M* in cell B8:D10.**

Select: **Cells F8:F10**

Enter: **The array formula =MMULT(MINVERSE(B8:D 10),F3:F5) in selected region. Use CRTL-SHIFT-ENTER to input this array formula.**

	A	B	C	D	E	F	G	H
1		Technology Matrix M				Final Demand		
2		A	E	M				
3	A	0.2	0	0.1		20		
4	E	0.4	0.2	0.1		10		
5	M	0	0.4	0.3		30		
6								
7		I-M				Output		
8		0.8	0	-0.1		=MMULT(MINVERSE(B8:D10),F3:F5)		
9		-0.4	0.8	-0.1				
10		0	-0.4	0.7				

The output matrix is found in cells **F8:F10**.

	A	B	C	D	E	F
1		Technology Matrix M				Final Demand
2		A	E	M		
3	A	0.2	0	0.1		20
4	E	0.4	0.2	0.1		10
5	M	0	0.4	0.3		30
6						
7		I-M				Output
8		0.8	0	-0.1		33
9		-0.4	0.8	-0.1		37
10		0	-0.4	0.7		64

Chapter 5

Section 5-3 Linear Programming in Two Dimensions: Geometric Approach

Example 2: Solving a Linear Programming Problem

Enter: The corner points of the feasible region S.

Enter: **Formula =3*A3+B3 in cell D3**

	A	B	C	D
1	Corner Point			
2	x	y		z=3x+y
3	3	6		=3*A3+B3
4	2	16		
5	8	4		

Copy: **Contents of cell D3 to cells D4:D5**

	A	B	C	D
1	Corner Point			
2	x	y		z=3x+y
3	3	6		15
4	2	16		22
5	8	4		28

Enter: **Formula =MIN(D3:D5) in cell D7**

Enter: **Formula =MAX(D3:D5) in cell D9**

	A	B	C	D	E
1	Corner Point				
2	x	y		z=3x+y	
3	3	6		15	
4	2	16		22	
5	8	4		28	
6	Min				
7				15	
8	Max				
9				=MAX(D3:D5)	

To find the corner point at which the minimum and/or maximum occur, use the **LOOKUP** function [obviously more useful for long lists of corner points].

Enter: Formula **=LOOKUP($D$7,$D$3:$D$5,A$3:A$5)** in cell **A7** [finds the *x*-value in column **A** that corresponds to the minimum value in column **D**; note the use of anchors ($) going forward]

	A	B	C	D	E
1	Corner Point				
2	x	y		z=3x+y	
3	3	6		15	
4	2	16		22	
5	8	4		28	
6	Min				
7	=LOOKUP($D7,$D$3:$D$5,A$3:A$5)				
8	Max				
9				28	

Copy: **Contents of cell A7 to cells B7**

Copy: **Contents of cell A7 to cells A9 and B9**

	A	B	C	D
1	Corner Point			
2	x	y		z=3x+y
3	3	6		15
4	2	16		22
5	8	4		28
6	Min			
7	3	6		15
8	Max			
9	8	4		28

Chapter 6

Section 6-2 Simplex Method: Maximization with Problem Constraints of the Form $\leq$

Example 1: Using the Simplex Method

Enter: The initial simplex tableau and identify the first pivot element.

	A	B	C	D	E	F	G
1		x_1	x_2	s_1	s_2	P	
2	s_1	4	1	1	0	0	28
3	s_2	2	3	0	1	0	24
4	P	-10	-5	0	0	1	0

Perform row operation: $\frac{1}{4}R_1 \to R_1$

Enter: **=1/4*B2 in cell B6**

Copy: **Contents of cell B6 to cells C6:G6**

Enter: **=B3 in cell B7**

Copy: **Contents of cell B7 to cells B7:G8**

	A	B	C	D	E	F	G
1		x_1	x_2	s_1	s_2	P	
2	s_1	4	1	1	0	0	28
3	s_2	2	3	0	1	0	24
4	P	-10	-5	0	0	1	0
5							
6	s_1	1	0.25	0.25	0	0	7
7	s_2						
8	P						

Perform row operations:
$(-2)R_1 + R_2 \to R_2$ and
$10R_1 + R_3 \to R_3$

Enter: **=B6 in cell B10**

Copy: **Contents of cell B10 to cells C10:G10**

Enter: **=(-2)*B6+B7 in cell B11**

	A	B	C	D	E	F	G
1		x_1	x_2	s_1	s_2	P	
2	s_1	4	1	1	0	0	28
3	s_2	2	3	0	1	0	24
4	P	-10	-5	0	0	1	0
5							
6	s_1	1	0.25	0.25	0	0	7
7	s_2	2	3	0	1	0	24
8	P	-10	-5	0	0	1	0
9							
10	x_1	1	0.25	0.25	0	0	7
11	s_2	=-2*B6+B7					
12	P						

Copy: **Contents of cell B11 to cells C11:G11**

Enter: **=10*B6+B8 in cell B12**

Copy: **Contents of cell B12 to cells C12:G12**

	A	B	C	D	E	F	G
1		x_1	x_2	s_1	s_2	P	
2	s_1	4	1	1	0	0	28
3	s_2	2	3	0	1	0	24
4	P	-10	-5	0	0	1	0
5							
6	s_1	1	0.25	0.25	0	0	7
7	s_2	2	3	0	1	0	24
8	P	-10	-5	0	0	1	0
9							
10	x_1	1	0.25	0.25	0	0	7
11	s_2	0	2.5	-0.5	1	0	10
12	P	=10*B6+B8					

To determine next pivot element, look above the negative element in the bottom row.

Enter: **Formula =G10/C10 in cell I10**

Copy: **Contents of cell I10 to cell I11**

	A	B	C	D	E	F	G	H	I
1		x_1	x_2	s_1	s_2	P			
2	s_1	4	1	1	0	0	28		
3	s_2	2	3	0	1	0	24		
4	P	-10	-5	0	0	1	0		
5									
6	s_1	1	0.25	0.25	0	0	7		
7	s_2	2	3	0	1	0	24		
8	P	-10	-5	0	0	1	0		
9									
10	x_1	1	0.25	0.25	0	0	7		=G10/C10
11	s_2	0	2.5	-0.5	1	0	10		
12	P	0	-2.5	2.5	0	1	70		

Perform row operation: $\dfrac{1}{2.5} R_2 \to R_2$

Enter: **=1/2.5*B11 in cell B15**

Copy: **Contents of cell B15 to cells C15:G15**

Enter: **=B10 in cell B14 and =B12 in cell B16**

Copy: **Contents of cell B14 to cells C14:G14 and the contents of cell B16 to cells C16:G16**

	A	B	C	D	E	F	G	H	I
1		x_1	x_2	s_1	s_2	P			
2	s_1	4	1	1	0	0	28		
3	s_2	2	3	0	1	0	24		
4	P	-10	-5	0	0	1	0		
5									
6	s_1	1	0.25	0.25	0	0	7		
7	s_2	2	3	0	1	0	24		
8	P	-10	-5	0	0	1	0		
9									
10	x_1	1	0.25	0.25	0	0	7		28
11	s_2	0	2.5	-0.5	1	0	10		4
12	P	0	-2.5	2.5	0	1	70		
13									
14	x_1	1	0.25	0.25	0	0	7		
15	s_2	0	1	-0.2	0.4	0	4		
16	P	0	-2.5	2.5	0	1	70		

Perform row operations:

$-0.25R_2 + R_1 \rightarrow R_1$ and

$2.5R_2 + R_3 \rightarrow R_3$

Enter: **=-0.25*B15+B14 in cell B18**

Copy: **Contents of cell B18 to cells C18:G18**

Enter: **=B15 in cell B19 and =2.5*B15+B16 in cell B20**

Copy: **Contents of cell B19 to cells C19:G19 and the contents of cell B20 to cells C20:G20**

	A	B	C	D	E	F	G
1		x_1	x_2	s_1	s_2	P	
2	s_1	4	1	1	0	0	28
3	s_2	2	3	0	1	0	24
4	P	-10	-5	0	0	1	0
5							
6	s_1	1	0.25	0.25	0	0	7
7	s_2	2	3	0	1	0	24
8	P	-10	-5	0	0	1	0
9							
10	x_1	1	0.25	0.25	0	0	7
11	s_2	0	2.5	-0.5	1	0	10
12	P	0	-2.5	2.5	0	1	70
13							
14	x_1	1	0.25	0.25	0	0	7
15	s_2	0	1	-0.2	0.4	0	4
16	P	0	-2.5	2.5	0	1	70
17							
18	x_1	1	0	0.3	-0.1	0	6
19	s_2	0	1	-0.2	0.4	0	4
20	P	=2.5*B15+B16					

Since all the indicators in the last row are nonnegative, the optimal solution has been found.

18	x_1	1	0	0.3	-0.1	0	6
19	s_2	0	1	-0.2	0.4	0	4
20	P	0	0	2	1	1	80

Example 1: Using the Simplex Method (alternative method using Excel's Solver)

Enter: **Any value (e.g. 0) in cell A2**

Enter: **Any value (e.g. 0) in cell B2**

Enter: **The formula =10*A2+5*B2 in cell C2** (representing the value of P)

	A	B	C	D
1	x_1	x_2	P	
2	0	0	=10*A2+5*B2	

Enter: **The formula =4*A2+B2 in cell A5** (representing the left hand side of inequality 1)

Enter: **The value 28 in cell B5**

Enter: **The formula =2*A2+3*B2 in cell A6** (representing the left hand side of inequality 2)

Enter: **The value 24 in cell B6**

	A	B	C
1	x_1	x_2	P
2	0	0	0
3			
4	LHS	RHS	
5	0	28	
6	=2*A2+3*B2		

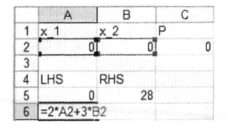

Choose: **Tools** > **Solver** >

Enter: Select Target Cell: **C2**

Select: Equal To: **Max**

Enter: By Changing Cells: **A2, B2**

Click: Subject to the Constraints:
Add

* If Solver does not appear, choose
Tools > Add-ins and select the Solver
add-in.

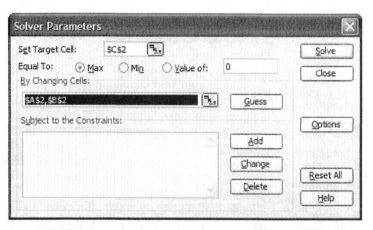

Enter: Cell Reference: **A5**

Select: <=

Enter: Constraint: **B5**

Click: **Add**

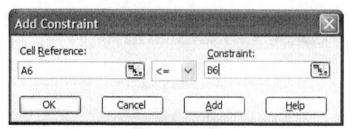

Enter: Cell Reference: **A6**

Select: <=

Enter: Constraint: **B6**

Click: **Add**

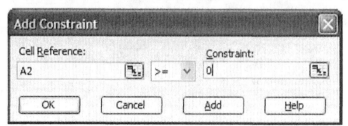

Enter: Cell Reference: **A2**

Select: >=

Enter: Constraint: **0**

Click: **Add**

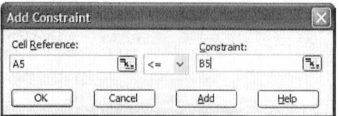

Enter: Cell Reference: **B2**

Select: >=

Enter: Constraint: **0**

Click: **OK**

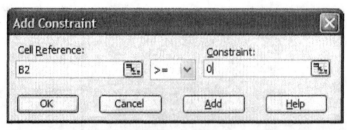

Click: **Solve**

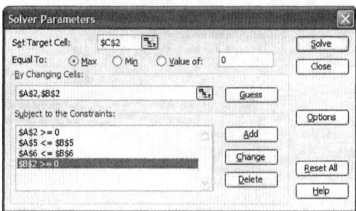

Select: **Keep Solver Solution**

Click: **OK**

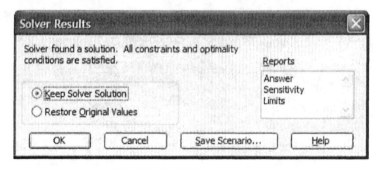

The optimal solution appears in cells **A2,B 2**, and the optimal value appears in cell **C2**.

	A	B	C
1	x_1	x_2	P
2	6	4	80
3			
4	LHS	RHS	
5	28	28	
6	24	24	

Example 2: Using the Simplex Method

Enter: **Any value (e.g. 0) in cell A2**

Enter: **Any value (e.g. 0) in cell B2**

Enter: **The formula =6*A2+3*B2 in cell C2** (representing the value of P)

Enter: **The formula =-2*A2+3*B2 in cell A5** (representing the left hand side of inequality 1)

Enter: **The value 9 in cell B5**

Enter: **The formula =-1*A2+3*B2 in cell A6** (representing the left hand side of inequality 2)

Enter: **The value 12 in cell B6**

Choose: **Tools > Solver >**

Enter: Select Target Cell: **C2**

Select: Equal To: **Max**

Enter: By Changing Cells: **A2, B2**

Click: Subject to the Constraints: **Add**

* If Solver does not appear, choose Tools > Add-ins and select the Solver add-in.

Enter: Cell Reference: **A5**

Select: **<=**

Enter: Constraint: **B5**

Click: **Add**

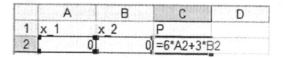

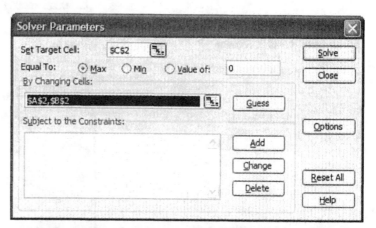

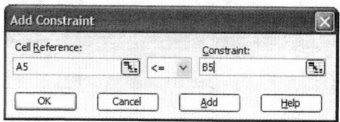

Enter: Cell Reference: **A6**

Select: <=

Enter: Constraint: **B6**

Click: **Add**

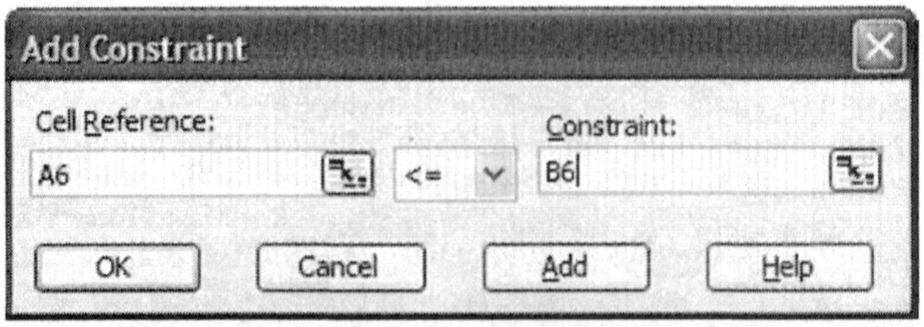

Add Constraint

Cell Reference:			Constraint:	
A6	📰	<= ∨	B6|	📰

OK	Cancel	Add	Help

Enter: Cell Reference: **A2**

Select: >=

Enter: Constraint: **0**

Click: **Add**

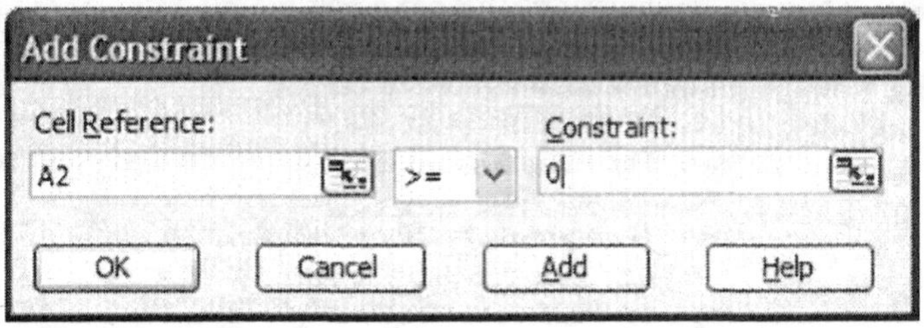

Add Constraint

Cell Reference:			Constraint:	
A2	📰	>= ∨	0|	📰

OK	Cancel	Add	Help

Enter: Cell Reference: **B2**

Select: >=

Enter: Constraint: **0**

Click: **OK**

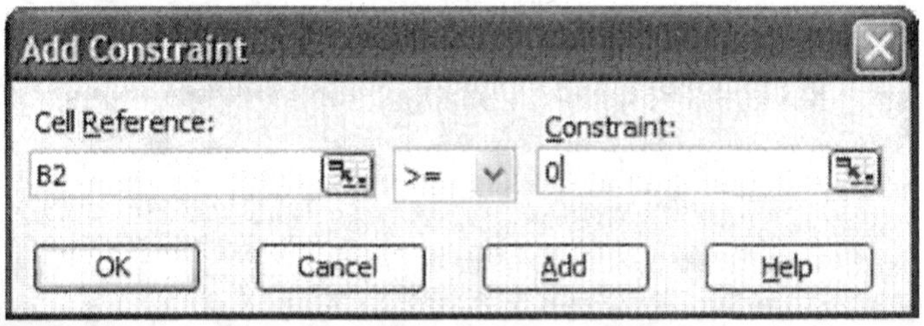

Add Constraint

Cell Reference:			Constraint:	
B2	📰	>= ∨	0|	📰

OK	Cancel	Add	Help

Click: **Solve**

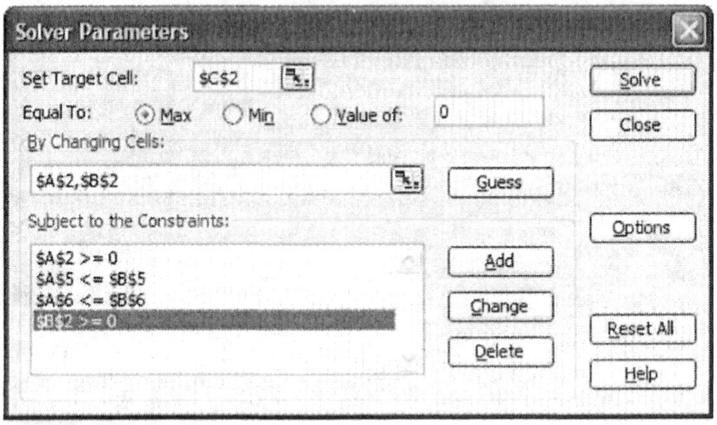

Solver Parameters

Set Target Cell: $C$2 📰

Equal To: ⦿ Max ○ Min ○ Value of: 0

By Changing Cells:

$A$2,$B$2 📰   Guess

Subject to the Constraints:

$A$2 >= 0
$A$5 <= $B$5
$A$6 <= $B$6
$B$2 >= 0

Add Change Delete

Solve Close Options Reset All Help

Select: **Keep Solver Solution**

Click: **OK**

Excel does not produce an optimal
solution (and rightfully so since no
optimal solution exists).

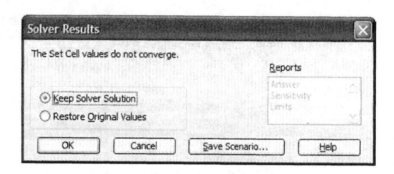

Example 3: Agriculture

Enter: **Data from problem in cells
B2:D4,E2 :E4, and B5:D5**

Enter: **The values 0 in cells B6:D6**
(representing a feasible solution)

Enter: **The formula
=B5*B6+C5*C6+D5*D6 in cell E5**
(representing the value of P)

	A	B	C	D	E	F	G
1	Resources	Crop A	Crop B	Crop C	Available	Used	
2	Acres	1	1	1	100		
3	Seed ($)	40	20	30	3200		
4	Workdays	1	2	1	160		
5	Profit Per Acre	100	300	200	=B5*B6+C5*C6+D5*D6		
6	Acres to plant	0	0	0		Profit	

Enter: **The formula =B2*$B$6+C2*$C$6+D2*$D$6 in cell F2** (representing the left hand side of inequality 1)

Copy: **The contents of cell F2 to cells F3:F4**

	A	B	C	D	E	F	G	H
1	Resources	Crop A	Crop B	Crop C	Available	Used		
2	Acres	1	1	1	100	=B2*$B$6+C2*$C$6+D2*$D$6		
3	Seed ($)	40	20	30	3200			
4	Workdays	1	2	1	160			
5	Profit Per Acre	100	300	200	0	<-Total		
6	Acres to plant	0	0	0		Profit		

Choose: **Tools > Solver >**

Enter: Select Target Cell: **E5**

Select: Equal To: **Max**

Enter: By Changing Cells: **B6:D6**

Click: Subject to the Constraints:
Add

* If Solver does not appear, choose
Tools > Add-ins and select the Solver
add-in.

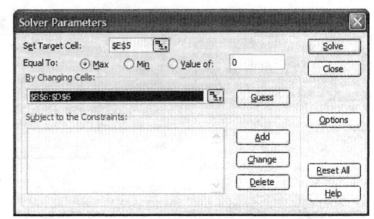

Enter: Cell Reference: **F2**

Select: <=

Enter: Constraint: **E2**

Click: **Add**

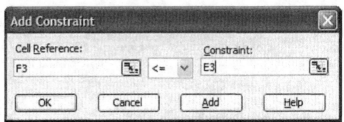

Enter: Cell Reference: **F3**

Select: <=

Enter: Constraint: **E3**

Click: **Add**

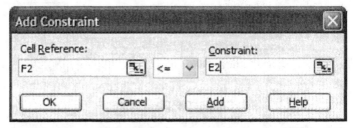

Enter: Cell Reference: **F4**

Select: <=

Enter: Constraint: **E4**

Click: **Add**

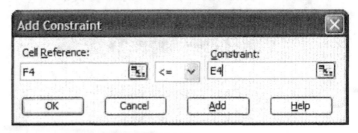

Enter: Cell Reference: **B6**

Select: >=

Enter: Constraint: **0**

Click: **Add**

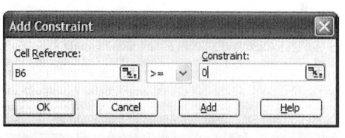

Enter: Cell Reference: **C6**

Select: >=

Enter: Constraint: **0**

Click: **Add**

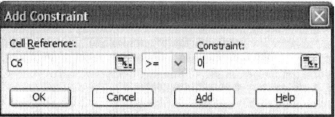

Enter: Cell Reference: **D6**

Select: >=

Enter: Constraint: **0**

Click: **OK**

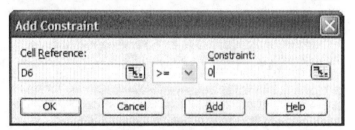

Click: **Solve**

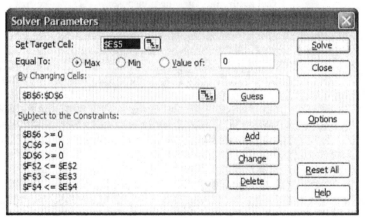

Select: **Keep Solver Solution**

Click: **OK**

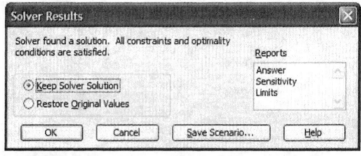

The optimal solution appears in cells **B6,C 6**, and **D6** and the optimal value appears in cell **E5**. Also, the amount of each resource used appears in column **F**.

	A	B	C	D	E	F
1	Resources	Crop A	Crop B	Crop C	Available	Used
2	Acres	1	1	1	100	100
3	Seed ($)	40	20	30	3200	2400
4	Workdays	1	2	1	160	160
5	Profit Per Acre	100	300	200	26000	<-Total
6	Acres to plant	0	60	40		Profit

Section 6-3 Dual Problem: Minimization with Problem Constraints of the Form ≥

Example 1: Forming the Dual Problem

Though easy to perform by hand, Excel easily computes the transpose of a matrix.

Select: **Cells A5:C8**

Enter: **The array formula =TRANSPOSE(A1:D3) in selected region.**

*Use CTRL-SHIFT-ENTER to input an array formula.

	A	B	C	D
1	2	1	5	20
2	4	1	1	30
3	40	12	40	1
4				
5	=TRANSPOSE(A1:D3)			
6				
7				
8				

Example 4: Transportation Problem

Enter: **Data from problem in cells B5:C6, D 5:D6, and B7:C7**

	A	B	C	D	E	F	G	H	I
1		Data					Shipping Schedule		
2		Distribution					Distribution		
3		Outlet		Assembly			Outlet		
4		I	II	Capacity			I	II	Total
5	Plant A	$6	$5	700		Plant A			
6	Plant B	$4	$8	900		Plant B			
7	Minimum	500	1000			Total			
8	Required						Total Cost		

Enter: **The values 0 in cells G5:H6**

Enter: **The formula =G5+H5 in cell I5**

Copy: **The formula in cell I5 to cell I6**

Enter: **The formula =G5+G6 in cell G7**

Copy: **The contents of cell G7 to cell H7**

Enter: **The formula =B5*G5+C5*H5+B6*G6+C6*H6 in cell I8** (representing the value of C)

	A	B	C	D	E	F	G	H	I	J	K
1		Data					Shipping Schedule				
2		Distribution					Distribution				
3		Outlet					Outlet				
4		I	II				I	II	Total		
5	Plant A	$6	$5	700		Plant A	0	0	0		
6	Plant B	$4	$8	900		Plant B	0	0	0		
7	Minimum	500	1000			Total	0	0			
8	Required						Total Cost		=B5*G5+C5*H5+B6*G6+C6*H6		

Enter: **The formula =G5+H5 in cell A12**

Enter: **The formula =D5 in cell B12**

Enter: **The formula =G6+H6 in cell A13**

Enter: **The formula =D6 in cell B13**

	A	B	C	D	E	F	G	H	I
1		Data					Shipping Schedule		
2		Distribution					Distribution		
3		Outlet		Assembly			Outlet		
4		I	II	Capacity			I	II	Total
5	Plant A	$6	$5	700		Plant A	0	0	0
6	Plant B	$4	$8	900		Plant B	0	0	0
7	Minimum	500	1000			Total	0	0	
8	Required						Total Cost		$0
9									
10	Constraints								
11	LHS	RHS							
12	0	700							
13	=G6+H6								

Enter: **The formula =G5+G6 in cell A14**

Enter: **The formula =B7 in cell B14**

Enter: **The formula =H5+H6 in cell A15**

Enter: The formula =C7 in cell B15

Choose: **Tools > Solver >**

Enter: Select Target Cell: **I8**

Select: Equal To: **Min**

Enter: By Changing Cells: **G5:H6**

Click: Subject to the Constraints:
Add

* If Solver does not appear, choose
Tools > Add-ins and select the Solver
add-in.

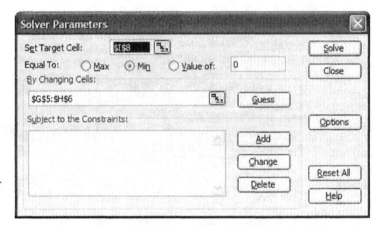

Enter: Cell Reference: **A12**

Select: <=

Enter: Constraint: **B12**

Click: **Add**

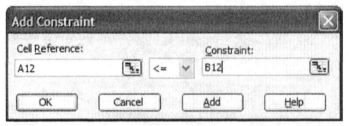

Repeat above step adding constraints:

A13 <= B13
A14 >= B14
A15 >= B15
G5 >= 0
H5 >= 0
G6 >= 0
H6 >= 0

Click: **OK**

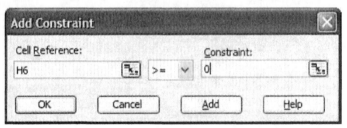

Click: **Solve**

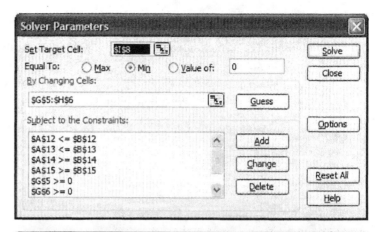

Select: **Keep Solver Solution**

Click: **OK**

The optimal solution appears in cells **G5,G6 ,H 5**, and **H6** and the optimal value appears in cell **I8**.

	A	B	C	D	E	F	G	H	I
1		Data					Shipping Schedule		
2		Distribution					Distribution		
3		Outlet		Assembly			Outlet		
4		I	II	Capacity			I	II	Total
5	Plant A	$6	$5	700		Plant A	0	700	700
6	Plant B	$4	$8	900		Plant B	500	300	800
7	Minimum	500	1000			Total	500	1000	
8	Required					Total Cost			$7,900
9									
10	Constraints								
11	LHS	RHS							
12	700	700							
13	800	900							
14	500	500							
15	1000	1000							

Section 6-4 Maximization and Minimization with Mixed Problem Constraints

Example 2: Using the Big *M* Method

Enter: **The preliminary simplex tableau for the modified problem.**

Enter: **Any relatively large number (compared to numbers in the tableau) in cell L1**

Enter: **The formula =$L$2 (absolute reference) in cells F5 and G5**

	A	B	C	D	E	F	G	H	I	J	K	L
1		x_1	x_2	x_3	s_1	a_1	s_2	a_2	P			1000
2	s_1	1	1	0	1	0	0	0	0	20		
3	a_1	1	0	1	0	1	0	0	0	5		
4	a_2	0	1	1	0	0	-1	1	0	10		
5		-1	1	-3	0	=$L$1						

Enter: **The formula =B2 in cell B7**

Copy: **Contents of cell B7 to cells B7:J9**

Enter: **The formula =-$L$1*B3+B5 in cell B10**

Copy: **Contents of cell B10 to cells C10:J10**

	A	B	C	D	E	F	G	H	I	J	K	L
1		x_1	x_2	x_3	s_1	a_1	s_2	a_2	P			1000
2	s_1	1	1	0	1	0	0	0	0	20		
3	a_1	1	0	1	0	1	0	0	0	5		
4	a_2	0	1	1	0	0	-1	1	0	10		
5		-1	1	-3	0	1000	0	1000	1	0		
6												
7	s_1	1	1	0	1	0	0	0	0	20		
8	a_1	1	0	1	0	1	0	0	0	5		
9	a_2	0	1	1	0	0	-1	1	0	10		
10		=-$L$1*B3+B5										

Enter: **The formula =B7 in cell B12**

Copy: **Contents of cell B12 to cells B12:J14**

Enter: **The formula =-$L$1*B9+B10 in cell B15**

Copy: **Contents of cell B15 to cells C15:J15**

	A	B	C	D	E	F	G	H	I	J	K	L
1		x_1	x_2	x_3	s_1	a_1	s_2	a_2	P			1000
2	s_1	1	1	0	1	0	0	0	0	20		
3	a_1	1	0	1	0	1	0	0	0	5		
4	a_2	0	1	1	0	0	-1	1	0	10		
5		-1	1	-3	0	1000	0	1000	1	0		
6												
7	s_1	1	1	0	1	0	0	0	0	20		
8	a_1	1	0	1	0	1	0	0	0	5		
9	a_2	0	1	1	0	0	-1	1	0	10		
10		-1001	1	-1003	0	0	0	1000	1	-5000		
11												
12	s_1	1	1	0	1	0	0	0	0	20		
13	a_1	1	0	1	0	1	0	0	0	5		
14	a_2	0	1	1	0	0	-1	1	0	10		
15		=-$L$1*B9+B10										

Pivot on the second row and third column. Perform the row operations $(-1) R_2 + R_3 \rightarrow R_3$ and $(2M+3) R_2 + R_4 \rightarrow R_4$.

Enter: **The formula =B12 in cell B17**

Copy: **Contents of cell B17 to cells B17:J18**

Enter: **The formula =-B13+B14 in cell B19**

Copy: **Contents of cell B19 to cells C19:J19**

Enter: **The formula =(2*$L$1+3)*B13+B15 in cell B20**

Copy: **Contents of cell B20 to cells C20:J20**

	A	B	C	D	E	F	G	H	I	J	K	L
1		x_1	x_2	x_3	s_1	a_1	s_2	a_2	P			1000
2	s_1	1	1	0	1	0	0	0	0	20		
3	a_1	1	0	1	0	1	0	0	0	5		
4	a_2	0	1	1	0	0	-1	1	0	10		
5		-1	1	-3	0	1000	0	1000	1	0		
6												
7	s_1	1	1	0	1	0	0	0	0	20		
8	a_1	1	0	1	0	1	0	0	0	5		
9	a_2	0	1	1	0	0	-1	1	0	10		
10		-1001	1	-1003	0	0	0	1000	1	-5000		
11												
12	s_1	1	1	0	1	0	0	0	0	20		
13	a_1	1	0	1	0	1	0	0	0	5		
14	a_2	0	1	1	0	0	-1	1	0	10		
15		-1001	-999	-2003	0	0	1000	0	1	-15000		
16												
17	s_1	1	1	0	1	0	0	0	0	20		
18	a_1	1	0	1	0	1	0	0	0	5		
19	a_2	-1	1	0	0	-1	-1	1	0	5		
20		=(2*$L$1+3)*B13+B15										

Pivot on the third row and second column. Perform the row operations $(-1)R_3 + R_1 \rightarrow R_1$ and $(M-1)R_3 + R_4 \rightarrow R_4$.

Enter: **The formula =B12 in cell B17**

Copy: **Contents of cell B17 to cells B17:J18**

Enter: **The formula =-B13+B14 in cell B19**

Copy: **Contents of cell B19 to cells C19:J19**

Enter: **The formula =(2*$L$1+3)*B13+B15 in cell B20**

Copy: **Contents of cell B20 to cells C20:J20**

	A	B	C	D	E	F	G	H	I	J	K	L
1		x_1	x_2	x_3	s_1	a_1	s_2	a_2	P			1000
2	s_1	1	1	0	1	0	0	0	0	20		
3	a_1	1	0	1	0	1	0	0	0	5		
4	a_2	0	1	1	0	0	-1	1	0	10		
5		-1	1	-3	0	1000	0	1000	1	0		
6												
7	s_1	1	1	0	1	0	0	0	0	20		
8	a_1	1	0	1	0	1	0	0	0	5		
9	a_2	0	1	1	0	0	-1	1	0	10		
10		-1001	1	-1003	0	0	0	1000	1	-5000		
11												
12	s_1	1	1	0	1	0	0	0	0	20		
13	a_1	1	0	1	0	1	0	0	0	5		
14	a_2	0	1	1	0	0	-1	1	0	10		
15		-1001	-999	-2003	0	0	1000	0	1	-15000		
16												
17	s_1	1	1	0	1	0	0	0	0	20		
18	x_3	1	0	1	0	1	0	0	0	5		
19	a_2	-1	1	0	0	-1	-1	1	0	5		
20		1002	-999	0	0	2003	1000	0	1	-4985		
21												
22	s_1	2	0	0	1	1	1	-1	0	15		
23	x_3	1	0	1	0	1	0	0	0	5		
24	x_2	-1	1	0	0	-1	-1	1	0	5		
25		3	0	0	0	1004	1	999	1	10		

Since the bottom row has no negative indicators, the optimal solution can be read from the tableau.

Example 3: Using the Big *M* Method

Enter: **Any value (e.g. 0) in cell A2**

Enter: **Any value (e.g. 0) in cell B2**

Enter: **The formula =3*A2+5*B2 in cell C2** (representing the value of P)

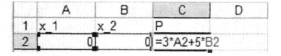

	A	B	C	D
1	x 1	x 2	P	
2	0	0	=3*A2+5*B2	

Enter: **The formula =2*A2+B2 in cell A5** (representing the left hand side of inequality 1)

Enter: **The value 4 in cell B5**

Enter: **The formula =A2+2*B2 in cell A6** (representing the left hand side of inequality 2)

Enter: **The value 10 in cell B6**

Choose: **Tools > Solver >**

Enter: Select Target Cell: **C2**

Select: Equal To: **Max**

Enter: By Changing Cells: **A2, B2**

Click: Subject to the Constraints: **Add**

* If Solver does not appear, choose Tools > Add-ins and select the Solver add-in.

Enter: Cell Reference: **A5**

Select: <=

Enter: Constraint: **B5**

Click: **Add**

Enter: Cell Reference: **A6**

Select: >=

Enter: Constraint: **B6**

Click: **Add**

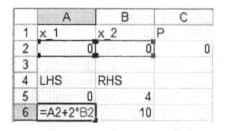

	A	B	C
1	x_1	x_2	P
2	0	0	0
3			
4	LHS	RHS	
5	0	4	
6	=A2+2*B2	10	

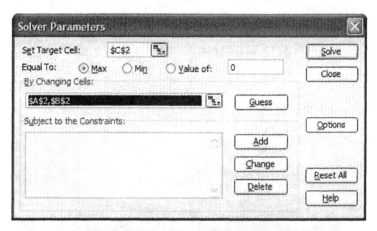

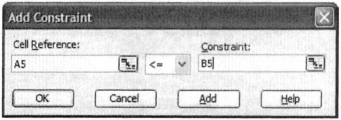

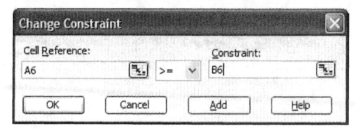

Enter: Cell Reference: **A2**

Select: >=

Enter: Constraint: **0**

Click: **Add**

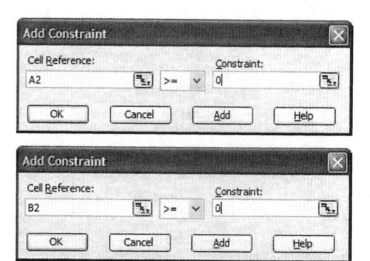

Enter: Cell Reference: **B2**

Select: >=

Enter: Constraint: **0**

Click: **OK**

Click: **Solve**

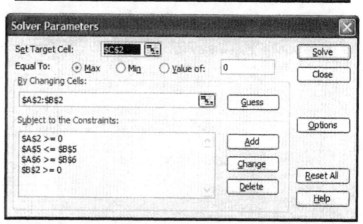

Select: **Keep Solver Solution**

Click: **OK**

Excel does not find a feasible
solution (and rightfully so since no
feasible solution exists).

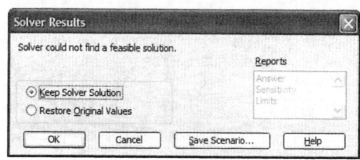

Chapter 7

Section 7-1 Logic

Example 1: Compound Propositions

Enter: **The value FALSE in cell A2**
(since proposition p is false)

Enter: **The value TRUE in cell B2**
(since proposition q is true)

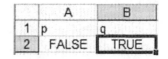

Enter: **The formula =NOT(A2) in cell A5**

Enter: **The formula =NOT(B2) in cell B5**

	A	B
1	p	q
2	FALSE	TRUE
3		
4	~p	~q
5	TRUE	=NOT(B2)

Enter: **The formula =OR(A2,B2) in cell C5**

	A	B	C	D	E
1	p	q			
2	FALSE	TRUE			
3					
4	~p	~q	p ∨ q	p ∧ q	p -> q
5	TRUE	FALSE	=OR(A2,B2)		

Enter: **The formula =AND(A2,B2) in cell D5**

	A	B	C	D	E
1	p	q			
2	FALSE	TRUE			
3					
4	~p	~q	p ∨ q	p ∧ q	p -> q
5	TRUE	FALSE	TRUE	=AND(A2,B2)	

Enter: **The formula =IF(A2,B2,TRUE) in cell D5**
(remember that if p is false, $p \to q$ is (vacuously) true)

	A	B	C	D	E	F
1	p	q				
2	FALSE	TRUE				
3						
4	~p	~q	p ∨ q	p ∧ q	p -> q	
5	TRUE	FALSE	TRUE	FALSE	=IF(A2,B2,TRUE)	

126

Example 2: Converse and Contrapositive

Enter: **The value TRUE in cell A2**
(since proposition *p* is true)

Enter: **The value FALSE in cell B2**
(since proposition *q* is false)

Enter: **The formula
=IF(A2,B2,TRUE) in cell A5**
(remember that if *p* is false, $p \rightarrow q$ is
(vacuously) true)

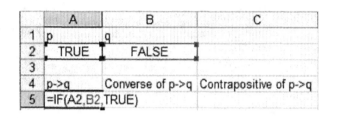

Enter: **The formula
=IF(B2,A2,TRUE) in cell B5**
(remember that if *q* is false, $q \rightarrow p$ is
(vacuously) true)

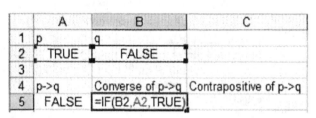

Enter: **The formula
=IF(NOT(B2),NOT(A2),TRUE) in
cell C5**

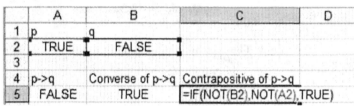

Example 3: Constructing Truth Tables

Enter: **All possible combinations of
variables *p* and *q* in cells A2:B5**

	A	B	C	D
1	p	q	~p	~p ∨ q
2	TRUE	TRUE		
3	TRUE	FALSE		
4	FALSE	TRUE		
5	FALSE	FALSE		

Enter: **The formula =NOT(A2) in
cell C2**

Copy: **The contents of cell C2 to
cells C3:C5**

	A	B	C	D
1	p	q	~p	~p ∨ q
2	TRUE	TRUE	FALSE	
3	TRUE	FALSE	FALSE	
4	FALSE	TRUE	TRUE	
5	FALSE	FALSE	TRUE	

127

Enter: **The formula =OR(C2,B2) in cell D2**

Copy: **The contents of cell D2 to cells D3:D5**

	A	B	C	D	E
1	p	q	~p	~p v q	
2	TRUE	TRUE	FALSE	=OR(C2,B2)	
3	TRUE	FALSE	FALSE		
4	FALSE	TRUE	TRUE		
5	FALSE	FALSE	TRUE		

Example 4: Constructing a Truth Table

Enter: **All possible combinations of variables** *p* **and** *q* **in cells A2:B5**

Enter: **The formula =IF(A2,B2,TRUE) in cell C2**

Copy: **The contents of cell C2 to cells C3:C5**

	A	B	C	D	E
1	p	q	p->q	(p->q) ^ p	[p->q) ^ p]->q
2	TRUE	TRUE	=IF(A2,B2,TRUE)		
3	TRUE	FALSE			
4	FALSE	TRUE			
5	FALSE	FALSE			

Enter: **The formula =AND(C2,A2) in cell D2**

Copy: **The contents of cell D2 to cells D3:D5**

	A	B	C	D	E
1	p	q	p->q	(p->q) ^ p	[p->q) ^ p]->q
2	TRUE	TRUE	TRUE	=AND(C2,A2)	
3	TRUE	FALSE	FALSE		
4	FALSE	TRUE	TRUE		
5	FALSE	FALSE	TRUE		

Enter: **The formula =IF(D2,B2,TRUE) in cell E2**

Copy: **The contents of cell E2 to cells E3:E5**

	A	B	C	D	E	F
1	p	q	p->q	(p->q) ^ p	[p->q) ^ p]->q	
2	TRUE	TRUE	TRUE	TRUE	=IF(D2,B2,TRUE)	
3	TRUE	FALSE	FALSE	FALSE		
4	FALSE	TRUE	TRUE	FALSE		
5	FALSE	FALSE	TRUE	FALSE		

Example 6: Verifying a Logical Implication

Enter: **All possible combinations of variables** *p* **and** *q* **in cells A2:B5**

Enter: **The formula =IF(A2,B2,TRUE) in cell C2**

Copy: **The contents of cell C2 to cells C3:C5**

	A	B	C	D	E	F
1	p	q	p->q	~q	(p->q) ^ ~q	~p
2	TRUE	TRUE	=IF(A2,B2,TRUE)			
3	TRUE	FALSE				
4	FALSE	TRUE				
5	FALSE	FALSE				

Enter: **The formula =NOT(B2) in cell D2**

Copy: **The contents of cell D2 to cells D3:D5**

	A	B	C	D	E	F
1	p	q	p->q	~q	(p->q) ^ ~q	~p
2	TRUE	TRUE	TRUE	=NOT(B2)		
3	TRUE	FALSE	FALSE			
4	FALSE	TRUE	TRUE			
5	FALSE	FALSE	TRUE			

Enter: **The formula =AND(C2,D2) in cell E2**

Copy: **The contents of cell E2 to cells E3:E5**

	A	B	C	D	E	F
1	p	q	p->q	~q	(p->q) ^ ~q	~p
2	TRUE	TRUE	TRUE	FALSE	=AND(C2,D2)	
3	TRUE	FALSE	FALSE	TRUE		
4	FALSE	TRUE	TRUE	FALSE		
5	FALSE	FALSE	TRUE	TRUE		

Enter: **The formula =NOT(A2) in cell E2**

Copy: **The contents of cell F2 to cells F3:F5**

	A	B	C	D	E	F
1	p	q	p->q	~q	(p->q) ^ ~q	~p
2	TRUE	TRUE	TRUE	FALSE	FALSE	FALSE
3	TRUE	FALSE	FALSE	TRUE	FALSE	FALSE
4	FALSE	TRUE	TRUE	FALSE	FALSE	TRUE
5	FALSE	FALSE	TRUE	TRUE	TRUE	TRUE

Section 7-4 Permutations and Combinations

Example 1: Computing Factorials

To compute 5! :

Enter: **The formula =FACT(5) in cell A1**

	A
1	=FACT(5)

To compute $\dfrac{7!}{6!}$:

Enter: **The formula =FACT(7)/FACT(6) in cell A1**

	A
1	=FACT(7)/FACT(6)

Example 3: Permutations

To compute $P_{13,8}$:

Enter: **The formula =PERMUT(13,8) in cell A1**

	A
1	=PERMUT(13,8)

Example 5: Combinations

To compute $C_{13,8}$:

Enter: **The formula =COMBIN(13,8) in cell A1**

	A
1	=COMBIN(13,8)

Chapter 8

Section 8-1 Sample Spaces, Events, and Probability

Example 6: Simulation and Empirical Probabilities

To simulate 100 rolls of two dice,

Enter: **The formula**
=ROUNDUP(6*RAND(),0) in cell A1

	A	B	C
1	=ROUNDUP(6*RAND(),0)		

The RAND function generates a random number between 0 and 1. Multiplying this by 6 has the effect of generating a random number between 0 and 6. Rounding up to 0 digits yields a random integer between 1 and 6.

Copy: **The contents of cell A1 to cells A1:B100**

Column **A** represents 100 rolls of one die and column **B** represents 100 rolls of the other.

	A	B
1	3	3
2	5	1
3	1	1
4	2	1
5	6	3
6	5	3
7	6	4
8	5	1

Enter: **The formula =SUM(A1:B1) in cell C1**

Copy: **The contents of cell C1 to cells C2:C100**

Column C represents the sum on the dice for each of the 100 rolls.

	A	B	C	D
1	3	3	=SUM(A1:B1)	
2	5	1		
3	1	1		
4	2	1		
5	6	3		
6	5	3		

Enter: **The value 2 in cell E2**

Enter: **The formula =E2+1 in cell E3**

Copy: **The contents of cell E3 to cells E4:E12**

Enter: **The formula =COUNTIF($C$1:$C$100,E2) in cell F2**

The COUNTIF function will count how many cells in column **C** have the same value as that found in cell **E2**

	A	B	C	D	E	F	G	H
1	6	4	10		Total	Frequency		
2	4	1	5		2	=COUNTIF($C$1:$C$100,E2)		
3	2	2	4		3			
4	3	5	8		4			
5	4	4	8		5			

Copy: **The contents of cell F2 to cells F3:F12**

	A	B	C	D	E	F
1	1	1	2		Total	Frequency
2	5	2	7		2	3
3	1	2	3		3	4
4	1	1	2		4	15
5	5	5	10		5	7
6	2	3	5		6	12
7	1	1	2		7	11
8	1	2	3		8	17
9	5	5	10		9	9
10	5	3	8		10	6
11	5	3	8		11	11
12	2	1	3		12	5

Choose: **Chart Wizard > Column >**

Select: **Chart sub-type: Clustered Column.**

Click: **Next**

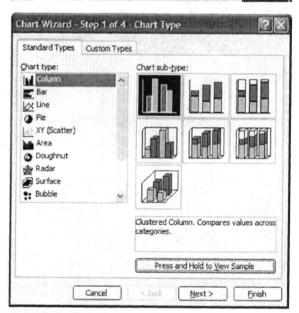

132

Enter: **Data Range F2:F12 or select cells**

Select: **Series in: columns**

Select: **Series tab**

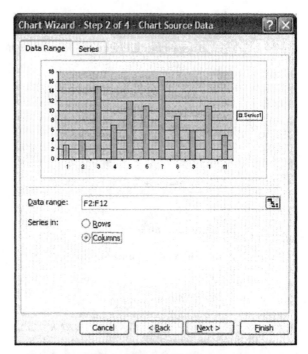

Select: Category (X) axis labels: by clicking on source data square at right of entry box.

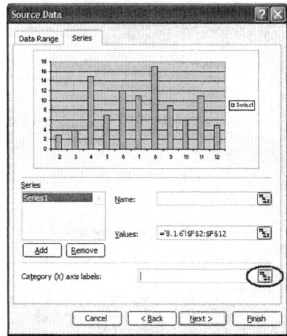

Select: **Cells E2:E12 using the mouse.**

Click: **Circled box on right of Source Data box**

Click: **Next**

	A	B	C	D	E	F	G
1	1	1	2		Total	Frequency	
2	5	2	7		2	3	
3	1	2	3		3	4	
4	1	1	2		4	15	
5	5	5	10		5	7	
6	2	3	5		6	12	
7	1	1	2		7	11	
8	1	2	3		8	17	
9	5	5	10		9	9	
10	5	3	8		10	6	
11	5	3	8		11	11	
12	2	1	3		12	5	

Source Data - Category (X) axis labels:

='8.1.6'!$E$2:$E$12

Select: **Titles Tab**

Enter: **Chart title, Value (X) axis, and Value (Y) axis**

Click: **Finish**

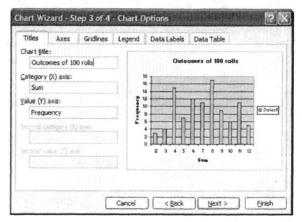

Chart Wizard - Step 3 of 4 - Chart Options

Tabs: Titles | Axes | Gridlines | Legend | Data Labels | Data Table

Chart title: Outcomes of 100 rolls

Category (X) axis: Sum

Value (Y) axis: Frequency

Cancel | < Back | Next > | Finish

Section 8-2 Union, Intersection, and Complement of Events; Odds

Example 5: Birthday Problem

To construct a table like that in Table 1,

Enter: **The value 5 in cell A2**

Enter: **The formula =A2+1 in cell A3**

Copy: **The contents of cell A3 to cells A4:A47**

Enter: **The formula =1-PERMUT(365,A2)/(365^A2) in cell B2**

	A	B	C	D
1	n	P(E)		
2	5	=1-PERMUT(365,A2)/(365^A2)		
3	6			
4	7			
5	8			

Copy: **The contents of cell C2 to cells C3:C47**

	A	B
1	n	P(E)
2	5	0.0271356
3	6	0.0404625
4	7	0.0562357
5	8	0.0743353
6	9	0.0946238
7	10	0.1169482
8	11	0.1411414
9	12	0.1670248
10	13	0.1944103
11	14	0.2231025
12	15	0.2529013

Section 8-5 Random Variable, Probability Distribution, and Expected Value

Example 2: Expected Value

Enter: **The values 0, 1, and 2 in cells B1:D1**

Enter: **The formula =COMBIN(2,B1)*COMBIN(18,3-B1)/COMBIN(20,3) in cell B2**

	A	B	C	D
1	x_i	0	1	2
2		=COMBIN(2,B1)*		
3		COMBIN(18,3-B1)/		
4		COMBIN(20,3)		

Copy: **The contents of cell B2 to cells C2:D2**

	A	B	C	D
1	x_i	0	1	2
2	p_i	0.715789	0.268421	0.015789

Enter: **The formula =B1*B2 in cell B3**

Copy: **The contents of cell B3 to cells C3:D3**

	A	B	C	D
1	x_i	0	1	2
2	p_i	0.715789	0.268421	0.015789
3		=B1*B2		

Enter: **The formula =SUM(B3:D5) in cell F5**

	A	B	C	D	E	F	G
1	x_i	0	1	2		E(X)	
2	p_i	0.715789	0.268421	0.015789			
3		0	0.268421	0.031579		=SUM(B3:D3)	

Chapter 9

Section 9-1 Properties of Markov Chains

Example 1: Insurance

Enter: **The transition matrix** P **in cells B1:C2**

Enter: **The initial-state matrix** S_0 **in cells B4:C4**

	A	B	C
1	P	0.23	0.77
2		0.11	0.89
3			
4	S_0	0.05	0.95

Find the first state S_1 :

Select: **Cells B6:C6**

Enter: **The array formula =MMULT(B4:C4,\$B\$1:\$C\$2) in selected region** (not in just a single cell).

*Use **CRTL-SHIFT-ENTER rather than ENTER to input an array formula.**

	A	B	C	D
1	P	0.23	0.77	
2		0.11	0.89	
3				
4	S_0	0.05	0.95	
5				
6	S_1	=MMULT(B4:C4,\$B\$1:\$C\$2)		

Find the second state S_2 :

Copy: **The contents of cells B6:C6 to cells B8:C8**

	A	B	C
1	P	0.23	0.77
2		0.11	0.89
3			
4	S_0	0.05	0.95
5			
6	S_1	0.116	0.884
7			
8	S_2	0.12392	0.87608

Example 2: Using P^k to Compute S_k

Enter: **The transition matrix** P **in cells B1:C2**

Select: **Cells B4:C5**

Enter: **The array formula =MMULT(B1:C2,B1:C2) in selected region**

*Use **CRTL-SHIFT-ENTER rather than ENTER to input an array formula.**

	A	B	C	D
1	P	0.1	0.9	
2		0.6	0.4	
3				
4	P^2	=MMULT(B1:C2,B1:C2)		
5				

Copy: **The contents of cells B4:C5 to cells B7:C8**

	A	B	C
1	P	0.1	0.9
2		0.6	0.4
3			
4	P^2	0.55	0.45
5		0.3	0.7
6			
7	P^4	0.4375	0.5625
8		0.375	0.625

Enter: **The initial-state matrix** S_0 **in cells E1:F1**

	A	B	C	D	E	F	G
1	P	0.1	0.9	S_0	0.2	0.8	
2		0.6	0.4				
3							
4	P^2	0.55	0.45				
5		0.3	0.7				
6							
7	P^4	0.4375	0.5625	S_4	=MMULT(E1:F1,B7:C8)		
8		0.375	0.625				

Select: **Cells E7:F7**

Enter: **The array formula =MMULT(E1:F1,B7:C8) in selected region**

*Use CRTL-SHIFT-ENTER rather than ENTER to input an array formula.

Example 4: University Enrollment

Enter: **The transition matrix** *P* **in cells B1:E4**

	A	B	C	D	E
1	P	0.6	0.1	0.3	0
2		0	1	0	0
3		0	0.1	0.5	0.4
4		0	0	0	1
5					
6	P^2	=MMULT(B1:E4,B1:E4)			
7					
8					
9					

Select: **Cells B6:E9**

Enter: **The array formula = MMULT(B1:E4,B1:E4) in selected region**

*Use CRTL-SHIFT-ENTER rather than ENTER to input an array formula.

Copy: **The contents of cells B6:E9 to cells B11:E14**

	A	B	C	D	E
1	P	0.6	0.1	0.3	0
2		0	1	0	0
3		0	0.1	0.5	0.4
4		0	0	0	1
5					
6	P^2	0.36	0.19	0.33	0.12
7		0	1	0	0
8		0	0.15	0.25	0.6
9		0	0	0	1
10					
11	P^4	0.1296	0.3079	0.2013	0.3612
12		0	1	0	0
13		0	0.1875	0.0625	0.75
14		0	0	0	1

Section 9-2 Regular Markov Chains

Example 1: Recognizing Regular Matrices

Enter: **The transition matrix *P* in cells B1:D3**

Select: **Cells B5:D7**

Enter: **The array formula =MMULT(B1:D3,$B$1:$D$3) in selected region**

*Use **CRTL-SHIFT-ENTER rather than ENTER to input an array formula.**

	A	B	C	D
1	P	0.5	0.5	0
2		0	0.5	0.5
3		1	0	0
4				
5	P^2	=MMULT(B1:D3,$B$1:$D$3)		
6				
7				

Copy: **The contents of cells B5:D7 to cells B9:D11**

*Note the use of the absolute reference in computing P^3

	A	B	C	D
1	P	0.5	0.5	0
2		0	0.5	0.5
3		1	0	0
4				
5	P^2	0.25	0.5	0.25
6		0.5	0.25	0.25
7		0.5	0.5	0
8				
9	P^3	0.375	0.375	0.25
10		0.5	0.375	0.125
11		0.25	0.5	0.25

Example 2: Finding the Stationary Matrix

Enter: **The transition matrix *P* in cells B1:C2**

Enter: **Any initial-state matrix *S* (e.g. [0.5 0.5]) in cells B4:C4**

Select: **Cells B6:C6**

Enter: **The array formula =MMULT(B4:C4,B1:C2) in selected region**

Use CRTL-SHIFT-ENTER rather than ENTER to input an array formula.

We seek S (having components which sum to 1) such that SP = S. To do this, we will minimize the difference between SP and S. That is, minimize the sum of the square of the differences between the components.

Enter: **The formula =SUM(B4:C4) in cell D4**

Enter: **The formula =B6-B4 in cell B8**

Copy: **The contents of cell B8 to cell C8**

	A	B	C
1	P	0.7	0.3
2		0.2	0.8
3			
4	S	0.5	0.5
5			
6	SP	=MMULT(B4:C4,B1:C2)	

	A	B	C	D	E
1	P	0.7	0.3		
2		0.2	0.8		
3					
4	S	0.5	0.5	=SUM(B4:C4)	
5					
6	SP	0.45	0.55		

	A	B	C	D
1	P	0.7	0.3	
2		0.2	0.8	
3				
4	S	0.5	0.5	1
5				
6	SP	0.45	0.55	
7				
8	Difference	-0.05	0.05	

Enter: **The formula =B8^2 in cell B9**

Copy: **The contents of cell B9 to cell C9**

Enter: **The formula =SUM(B9:C9) in cell D9**

	A	B	C	D
1	P	0.7	0.3	
2		0.2	0.8	
3				
4	S	0.5	0.5	1
5				
6	SP	0.45	0.55	
7				
8	Difference	-0.05	0.05	
9	Diff Sq	=B8^2		

Choose: **Tools > Solver >**

Enter: Select Target Cell: **D9**

Select: Equal To: **Min**

Enter: By Changing Cells: **B4:C4**

Click: Subject to the Constraints: **Add**

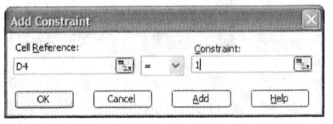

Enter: Cell Reference: **D4**

Select: **=**

Enter: Constraint: **1**

Click: **OK**

Click: **Solve**

Select: **Keep Solver Solution**

Click: **OK**

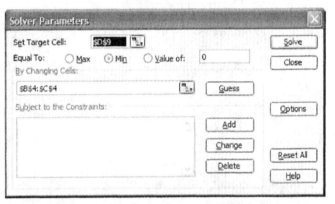

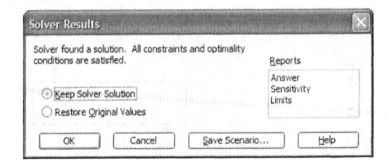

Since exactly one solution exists, an
approximation to it appears in cells
B4 and **C4**.

	A	B	C	D
1	P	0.7	0.3	
2		0.2	0.8	
3				
4	S	0.4	0.6	1
5				
6	SP	0.4	0.6	
7				
8	Difference	0	0	
9	Diff Sq	0	0	0

Theorem 1 allows a different
approach:

Enter: **The transition matrix *P* in
cells B1:C2**

Select: **Cells B4:C5**

Enter: **The array formula
=MMULT(B1:C2,$B$1:$C$2) in
selected region**

*Use **CRTL-SHIFT-ENTER** rather
than **ENTER** to input an array
formula.

	A	B	C	D
1	P	0.7	0.3	
2		0.2	0.8	
3				
4	P^2	=MMULT(B1:C2,$B$1:$C$2)		
5				

Copy: **The contents of cells B4:C5 to cells B7:C8**

Repeat as necessary. The resulting products will approach a matrix having rows consisting of the stationary matrix.

	A	B	C
1	P	0.7	0.3
2		0.2	0.8
3			
4	P^2	0.55	0.45
5		0.3	0.7
6			
7	P^3	0.475	0.525
8		0.35	0.65
9			
10	P^4	0.4375	0.5625
11		0.375	0.625
12			
13	P^5	0.41875	0.58125
14		0.3875	0.6125
15			
16	P^6	0.409375	0.590625
17		0.39375	0.60625

Example 5: Approximating the Stationary Matrix

Enter: **The transition matrix *P* in cells B1:D3**

Select: **Cells B5:D7**

Enter: **The array formula =MMULT(B1:D3,B1:D3) in selected region**

*Use CRTL-SHIFT-ENTER rather than ENTER to input an array formula.

	A	B	C	D
1	P	0.5	0.2	0.3
2		0.7	0.1	0.2
3		0.4	0.1	0.5
4				
5	P^2	=MMULT(B1:D3,B1:D3)		
6				
7				

Copy: **The contents of cells B5:D7 to cells B9:D11**

Repeat as necessary to approximate $\overline{P}$

	A	B	C	D
1	P	0.5	0.2	0.3
2		0.7	0.1	0.2
3		0.4	0.1	0.5
4				
5	P^2	0.51	0.15	0.34
6		0.5	0.17	0.33
7		0.47	0.14	0.39
8				
9	P^4	0.4949	0.1496	0.3555
10		0.4951	0.1501	0.3548
11		0.493	0.1489	0.3581
12				
13	P^8	0.494254	0.149426	0.35632
14		0.494256	0.149427	0.356317
15		0.494249	0.149424	0.356327

Chapter 10

Section 10-2 Mixed Strategy Games

Example 1: Solving a 2×2 **Nonstrictly Determined Matrix Game**

Enter: **The game matrix** M **in cells A2:B3**

Enter: **The formula =A2+B3-B2-A3 for** D **in cell D2**

Enter: **The formula =(B3-A3)/D2 in cell A6**

Enter: **The formula =(A2-B2)/D2 in cell B6**

Enter: **The formula =(B3-B2)/D2 in cell A9**

Enter: **The formula =(A2-A3)/D2 in cell A10**

Enter: **The formula =(A2*B3-B2*A3)/D2 in cell A6**

*Note: This setup will solve all 2×2 nonstrictly determined matrix games – simply change the values in cells **A2:B3**.

	A	B	C	D	E
1	M			D	
2	2	-3		=A2+B3-B2-A3	
3	-3	4			
4					
5	P*				
6	0.583333	0.416667			
7					
8	Q*				
9	0.583333				
10	0.416667				
11					
12	v				
13	-0.08333				

Section 10-3 Linear Programming and 2×2 Games: Geometric Approach

Example 1: Solving 2×2 **Matrix Games Using Geometric Methods**

Enter: **The game matrix** M **in cells A2:B3**

Convert M to a positive matrix:

Enter: **The formula =A2+4 in cell D2**

Copy: **The contents of cell D2 to cells D2:E3**

	A	B	C	D	E
1	M			M 1	
2	-2	4		=A2+4	
3	1	-3			

144

Enter: **Any values (e.g. 0) in cells A6 and B6** (representing the values of x_1 and x_2)

Enter: **The formula =A6+B6 in cell D6** (representing the value of y)

Enter: **The formula =D2*A6+D3*B6 in cell A9**

Enter: **The value 1 in cell B9**

Enter: **The formula =E2*A6+E3*B6 in cell A10**

Enter: **The value 1 in cell B10**

Choose: **Tools > Solver >**

Enter: Select Target Cell: **D6**

Select: Equal To: **Min**

Enter: By Changing Cells: **A6:B6**

Click: Subject to the Constraints: **Add**

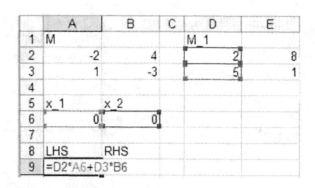

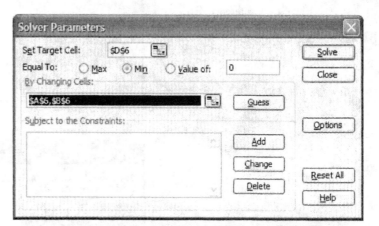

Enter: Cell Reference: **A9**

Select: **>=**

Enter: Constraint: **B9**

Click: **Add**

Click: **Solve**

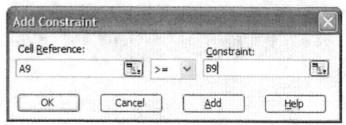

Enter: Cell Reference: **A10**

Select: >=

Enter: Constraint: **B10**

Click: **Add**

Click: **Solve**

Enter: Cell Reference: **A6**

Select: >=

Enter: Constraint: **0**

Click: **Add**

Enter: Cell Reference: **B6**

Select: >=

Enter: Constraint: **0**

Click: **OK**

Click: **Solve**

Select: **Keep Solver Solution**

Click: **OK**

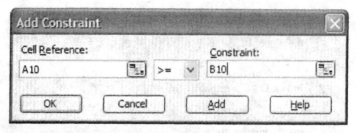

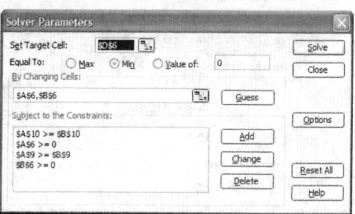

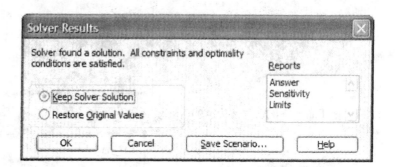

The solution will appear in cells **A6** and **B6**.

	A	B	C	D	E
1	M			M_1	
2	-2	4		2	8
3	1	-3		5	1
4					
5	x_1	x_2		y	
6	0.105263	0.157895		0.263158	
7					
8	LHS	RHS			
9	1	1			
10	1	1			

Repeat the above steps on the corresponding maximization problem.

	A	B	C	D	E
1	M			M_1	
2	-2	4		2	8
3	1	-3		5	1
4					
5	x_1	x_2		y	
6	0.105263	0.157895		0.263158	
7					
8	LHS	RHS			
9	1	1			
10	1	1			
11					
12	z_1	z_2		y	
13	0	0		0	
14					
15	LHS	RHS			
16	=D2*A13+E2*B13				

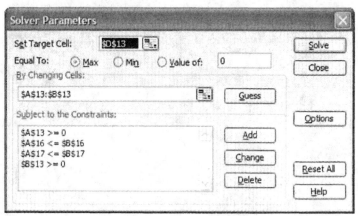

Solver Parameters

Set Target Cell: $D$13

Equal To: ⦿ Max ◯ Min ◯ Value of: 0

By Changing Cells:

$A$13:$B$13

Subject to the Constraints:

$A$13 >= 0
$A$16 <= $B$16
$A$17 <= $B$17
$B$13 >= 0

Solve   Close   Guess   Options   Add   Change   Delete   Reset All   Help

Enter: **The formula =1/D6 in cell B19** (representing the value of v_1)

Enter: **The formula =B19*A6 in cell A22**

Enter: **The formula =B19*B6 in cell B22**

Enter: **The formula =B19*A13 in cell D21**

Enter: **The formula =B19*B13 in cell D22**

*Note: The value of the original game is 4 less than the value in cell **B19**

	A	B	C	D	E
1	M			M_1	
2	-2	4		2	8
3	1	-3		5	1
4					
5	x_1	x_2		y	
6	0.105263	0.157895		0.263158	
7					
8	LHS	RHS			
9	1	1			
10	1	1			
11					
12	z_1	z_2		y	
13	0.184211	0.078947		0.263158	
14					
15	LHS	RHS			
16	1	1			
17	1	1			
18					
19	v_1	3.8			
20				Q*	
21	P*			0.7	
22	0.4	0.6		=B19*B13	

Chapter 11

Section 11-1 Graphing Data

Figure 1: Creating a vertical bar graph

Enter: Data from Table 1

	A	B
1	Year	Debt (billions $)
2	1960	284.1
3	1970	370.1
4	1980	907.7
5	1990	3233.3
6	2000	5674.2

Choose: **Chart Wizard > Column >**

Select: **Chart sub-type: Clustered Column.**

Click: **Next**

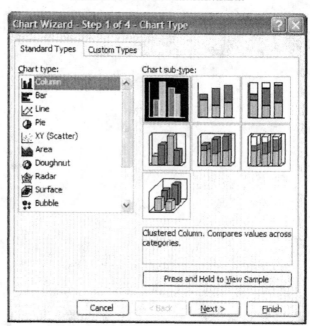

Enter: Data range: **B2:B6**

Select: **Series in: Columns**

Click: **Series Tab**

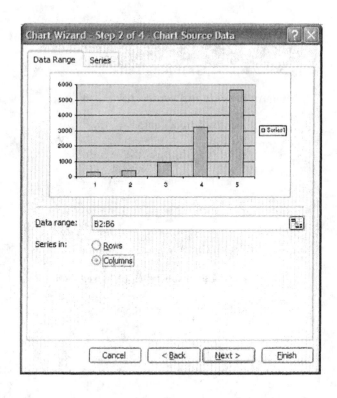

Enter: Name: **U.S. Public Debt**

Select: Category (X) axis labels: by clicking on source data square at right of entry box.

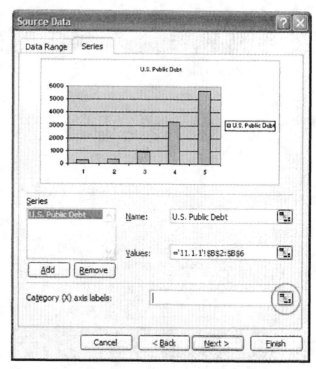

Select: **Cells A2:A6 using the mouse.**

Click: **Circled box on right of Source Data box**

Click: **Next**

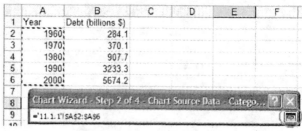

Enter: Category (X) axis: **Year**

Enter: Value (Y) axis: **Billion dollars**

Click: **Finish**

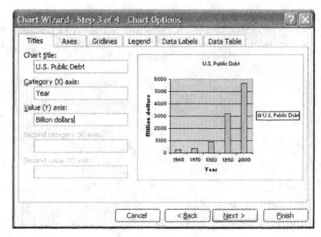

Several additional options are available in formatting the bar graph.

On any bar in the graph,

Right-click and select: **Format Data Series...**

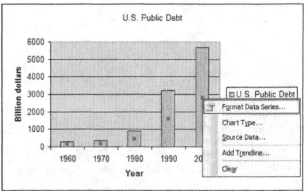

As an example, the bars can be widened.

Select: **Options Tab**

Enter: **Gap width: 50**

Click: **OK**

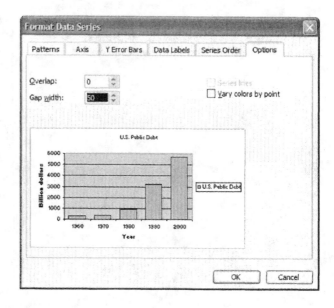

Figure 2: Creating a horizontal bar graph

Enter: Data from Table 2

	A	B
1	Airport	Arrivals and Departures
2	Atlanta	86
3	Chicago (O'Hare)	77
4	Los Angeles	61
5	Dallas/Ft. Worth	59
6	Las Vegas	44

Choose: **Chart Wizard > Bar >**

Select: **Chart sub-type: Clustered Bar.**

Click: **Next**

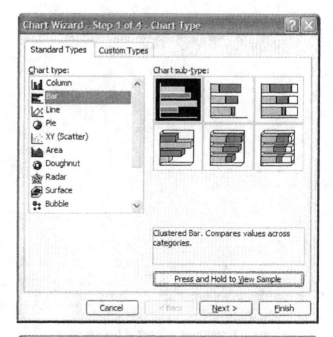

Enter: Data range: **B2:B6** (or highlight data using mouse)

Select: **Series in: Columns**

Click: **Series Tab**

Enter: Name: **Traffic at Busiest U.S. Airports, 2005**

Select: Category (X) axis labels: by clicking on source data square at right of entry box.

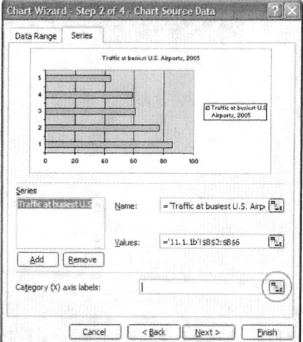

Select: **Cells A2:A6 using the mouse.**

Click: **Circled box on right of Source Data box**

Click: **Next**

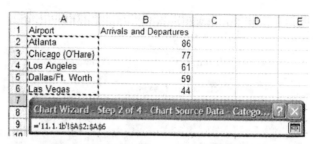

Click: **Titles Tab**

Enter: Value (Y) axis: **Arrivals and departures (million passengers)**

Click: **Legend Tab**

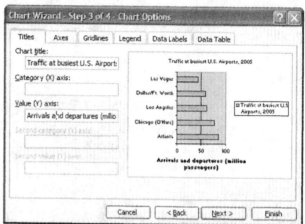

De-select: **Show legend**

Click: **Finish**

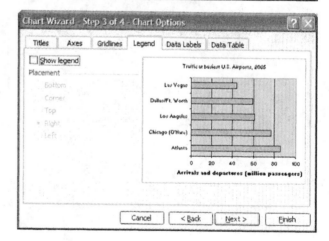

To arrange the categories as shown in Figure 2 in the text,

Right-click: On the vertical (category) axis

Select: **Format Axis...**

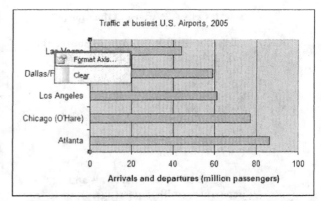

Select: **Scale Tab**

Select: **Categories in reverse order**

Select: **Value (Y) axis crosses at maximum category**

Click: **OK**

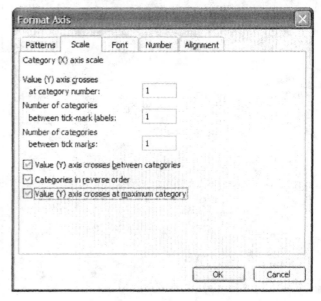

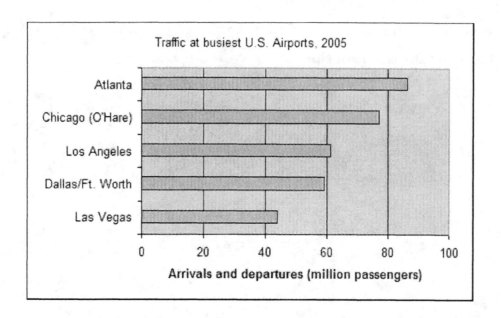

Figure 3: Creating a double bar graph

Enter: Data for the graph

	A	B	C
1	Education	Male Income	Female Income
2	Bachelor's degree	$56,502	$41,327
3	Associate degree	$42,871	$32,253
4	Some college	$41,348	$30,142
5	High school diploma	$35,412	$26,074
6	Some high school	$26,468	$18,938

Choose: **Chart Wizard > Bar >**

Select: **Chart sub-type: Clustered Bar.**

Click: **Next**

Select: **Data Range Tab**

Enter: Data range: **B2:C6** (or highlight data using mouse)

Select: **Series in: Columns**

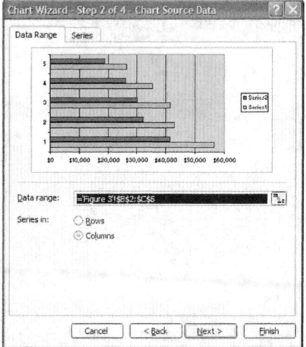

Click: **Series Tab**

Select: Series: **Series 1**

Enter: Name: **Male**

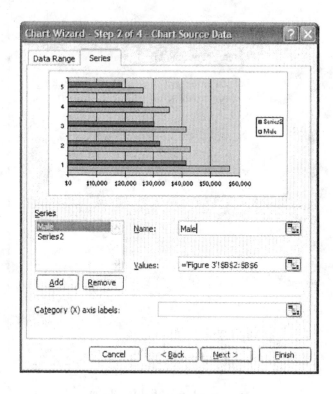

Select: Series: **Series 2**

Enter: Name: **Female**

Select: Category (X) axis labels: by clicking on source data square at right of entry box.

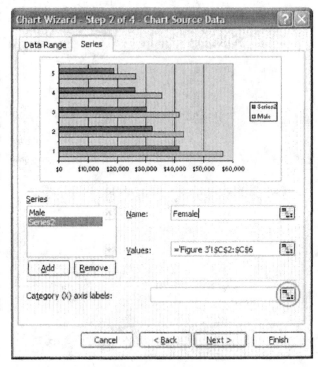

Select: **Cells A2:A6 using the mouse.**

Click: **Circled box on right of Source Data box**

Click: **Next**

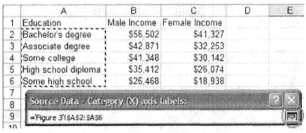

	A	B	C	D	E
1	Education	Male Income	Female Income		
2	Bachelor's degree	$56,502	$41,327		
3	Associate degree	$42,871	$32,253		
4	Some college	$41,348	$30,142		
5	High school diploma	$35,412	$26,074		
6	Some high school	$26,468	$18,938		
7					
8	Source Data - Category (X) axis labels:				
9	='Figure 3'!$A$2:$A$6				
10					

Select: Titles Tab

Enter: Chart title: **Education and Income, 2003**

Enter: Value (Y) axis: **Median annual income ($)**

Select: **Legend Tab**

De-select: **Show legend**

Click: **Finish**

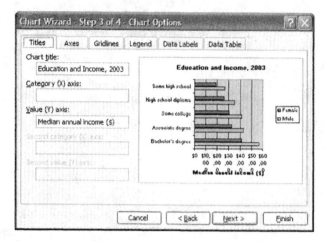

Select: Top bar of 'Some high school' (click twice)

Right-click and Select: **Format Data Point...**

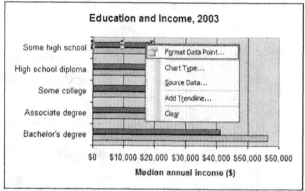

Select: **Data Labels Tab**

Select: Label Contains: **Series name**

Click: **OK**

Repeat with bottom bar of 'Some high school'

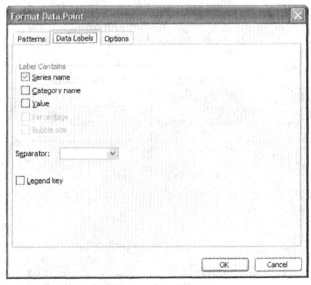

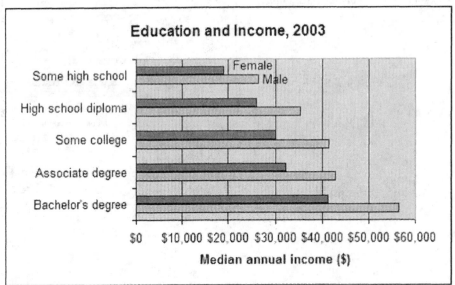

Figure 5: Creating a broken-line graph

Create the graph shown in Figure 1.

Select: **The chart** (click on the chart).

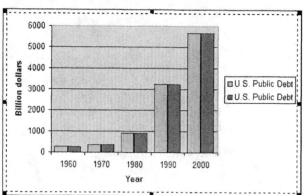

Copy the chart onto itself.

Enter: **CTRL-C** followed by **CTRL-V**

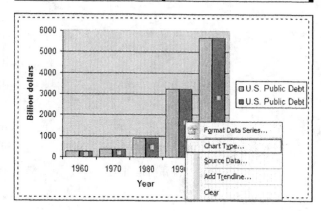

Right Click: Second set of bars

Select: **Chart Type…**

Select: Chart type: **Line**

Select: Chart sub-type: **Line with markers displayed at each data value.**

Click: **OK**

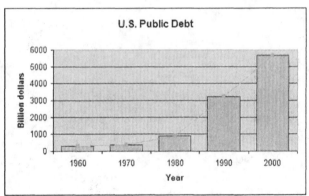

If necessary, add title by right clicking and selecting Chart Options…

If necessary, delete legend.

Figure 8: Creating a pie graph

Enter: Data for the graph

	A	B
1	Army	485536
2	Navy	384576
3	Air Force	369721
4	Marine Corps	173385
5	Coast Guard	37166

Choose: **Chart Wizard > Pie >**

Select: **Chart sub-type: Pie.**

Click: **Next**

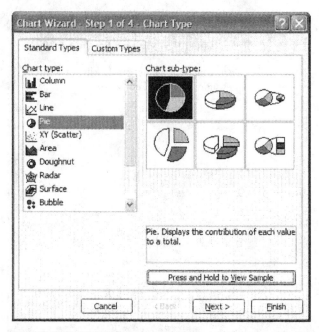

Select: **Data Range Tab**

Enter: Data range: **B1:B5** (or highlight these cells)

Select: Series in: **Columns**

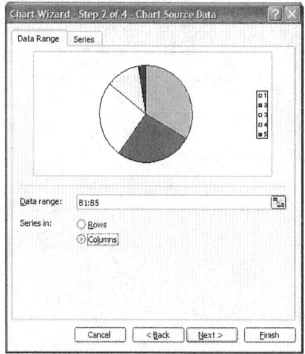

Select: **Series Tab**

Enter: Name: **U.S. Military Personnel, 2002**

Select: Category labels: by clicking on source data square at right of entry box.

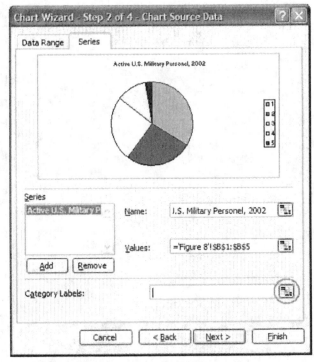

Select: **Cells A1:A5 using the mouse.**

Click: **Circled box on right of Source Data box**

Click: **Next**

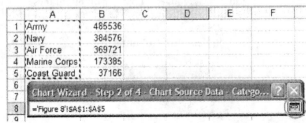

Select: **Legend Tab**

De-select: **Show Legend**

Select **Data Labels Tab**

Select: Label Contains: **Category name**

Select: Label Contains: **Percentage**

Click: **Finish**

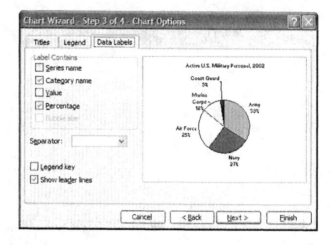

Figure 11: Creating a frequency table (and histogram)

Enter: Data (from Table 3) in cells A1:A100

	A
1	762
2	433
3	712
4	566
5	618
6	340
7	548
8	442
9	663
10	544
11	451
12	508

Set up the class intervals.

Enter: **The value 299.5 in cell C2 and the value 349.5 in cell D2**

Enter: **The formula =D2 in cell C3 and the formula =D2+50 in cell D3**

	A	B	C	D
1	762		Class Interval	
2	433		299.5	349.5
3	712		349.5	=D2+50

Copy: **The contents of cells C3:D3 to cells C4:D11**

	A	B	C	D
1	762		Class Interval	
2	433		299.5	349.5
3	712		349.5	399.5
4	566		399.5	449.5
5	618		449.5	499.5
6	340		499.5	549.5
7	548		549.5	599.5
8	442		599.5	649.5
9	663		649.5	699.5
10	544		699.5	749.5
11	451		749.5	799.5

Enter: **The formula =COUNTIF($A$1:$A$100,"<="&D2)-COUNTIF($A$1,$A$100,"<"&C2) in cell E2**

	A	B	C	D	E	F	G	H	I	J
1	762		Class Interval		Frequency					
2	433		299.5	349.5	=COUNTIF($A$1:$A$100,"<="&D2)-COUNTIF($A$1:$A$100,"<"&C2)					

Copy: **The contents of cell E2 to cells E3:E11**

	A	B	C	D	E
1	762		Class Interval		Frequency
2	433		299.5	349.5	1
3	712		349.5	399.5	2
4	566		399.5	449.5	5
5	618		449.5	499.5	10
6	340		499.5	549.5	21
7	548		549.5	599.5	20
8	442		599.5	649.5	19
9	663		649.5	699.5	11
10	544		699.5	749.5	7
11	451		749.5	799.5	4

Note: Frequency tables and histograms can be constructed using the Histogram option on the Data Analysis tool (You can find/add this tool under Add-ins on the Tool menu). However, Excel does a sloppy job of properly labeling the *x*-axis and a few tricks are necessary to deal with this problem. A simple search of the internet will reveal the (cumbersome!) tricks that can be used to create a histogram such as that shown in Figure 11.

Figure 13: Creating a frequency polygon

Construct a frequency table in the same manner as described above. Add a class interval to the left and right (both having frequency of 0).

	A	B	C	D	E
1	762		Class Interval		Frequency
2	433		249.5	299.5	0
3	712		299.5	349.5	1
4	566		349.5	399.5	2
5	618		399.5	449.5	5
6	340		449.5	499.5	10
7	548		499.5	549.5	21
8	442		549.5	599.5	20
9	663		599.5	649.5	19
10	544		649.5	699.5	11
11	451		699.5	749.5	7
12	508		749.5	799.5	4
13	415		799.5	849.5	0
14	493				
15	581				

Insert new column between D and E.

Highlight existing column E (left click on E).

Choose: **Insert > Columns**

	A	B	C	D	E	F
1	762		Class Interval			quency
2	433		249.5	299.5		0
3	712		299.5	349.5		1
4	566		349.5	399.5		2
5	618		399.5	449.5		5
6	340		449.5	499.5		10

Enter: **The formula =(C2+D2)/2 in cell E2**

	A	B	C	D	E	F
1	762		Class Interval		Midpoint	Frequency
2	433		249.5	299.5	=(C2+D2)/2	

Copy: **The contents of cell E2 to cells E3:E11**

	A	B	C	D	E	F
1	762		Class Interval		Midpoint	Frequency
2	433		249.5	299.5	274.5	0
3	712		299.5	349.5	324.5	1
4	566		349.5	399.5	374.5	2
5	618		399.5	449.5	424.5	5
6	340		449.5	499.5	474.5	10
7	548		499.5	549.5	524.5	21
8	442		549.5	599.5	574.5	20
9	663		599.5	649.5	624.5	19
10	544		649.5	699.5	674.5	11
11	451		699.5	749.5	724.5	7
12	508		749.5	799.5	774.5	4
13	415		799.5	849.5	824.5	0

Choose: **Chart Wizard > XY (Scatter) >**

Select: **Chart sub-type: Scatter with data points connected by lines.**

Click: **Next**

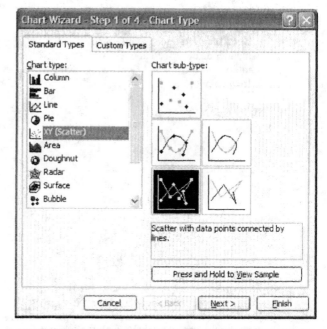

Select: **Data Range Tab**

Enter: Data range: **F2:F13**

Select: Series in: **Columns**

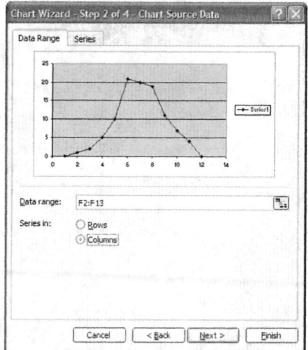

Select: **Series Tab**

Enter: Name: **Entrance Examination Scores**

Select: X Values: by clicking on source data square at right of entry box.

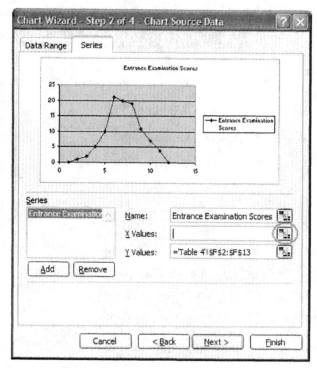

Select: **Cells E2:E13 using the mouse.**

Click: **Circled box on right of Source Data box**

Click: **Next**

	A	B	C	D	E	F	G
1	762		Class Interval		Midpoint	Frequency	
2	433		249.5	299.5	274.5	0	
3	712		299.5	349.5	324.5	1	
4	566		349.5	399.5	374.5	2	
5	618		399.5	449.5	424.5	5	
6	340		449.5	499.5	474.5	10	
7	548		499.5	549.5	524.5	21	
8	442		549.5	599.5	574.5	20	
9	663		599.5	649.5	624.5	19	
10	544		649.5	699.5	674.5	11	
11	451		699.5	749.5	724.5	7	
12	508		749.5	799.5	774.5	4	
13	415		799.5	849.5	824.5	0	
14							

Chart Wizard - Step 2 of 4 - Chart Source Data - X Valu... [?] [X]

='Table 4'!$E$2:$E$13

Select: **Titles Tab**

Enter: Value (X) axis: **Entrance examination scores**

Enter: **Value (Y) axis: Frequency**

Select: **Legend Tab**

De-select: **Show Legend**

Click: **Finish**

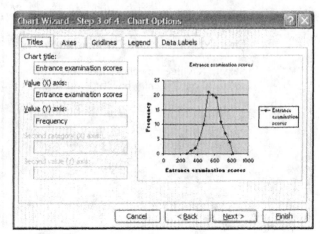

Right-click: on *x*-axis

Select: **Format axis...**

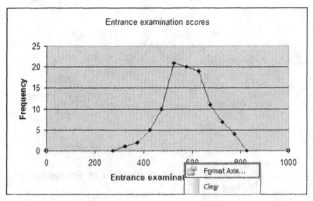

Select: **Scale Tab**

Enter: Minimum: **274.5**

Enter: Maximum: **824.5**

Enter: Major unit: **50**

Enter: Minor unit: **50**

Click: **OK**

Optional:

Right-click: on chart

Select: **Chart Options**

Select: **Gridlines Tab**

Select: Value (X) axis: **Major gridlines**

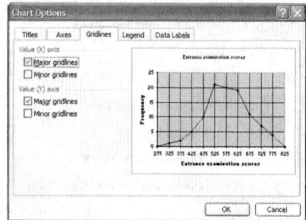

Figure 14: Creating a cumulative frequency polygon (table)

Start with the frequency table constructed above.

Enter: **The formula =SUM($F$2:F2) to cell G2**

	A	B	C	D	E	F	G
1	762		Class Interval		Midpoint	Frequency	Cumulative Frequency
2	433		249.5	299.5	274.5	0	=SUM($F$2:F2)
3	712		299.5	349.5	324.5	1	
4	566		349.5	399.5	374.5	2	
5	618		399.5	449.5	424.5	5	
6	340		449.5	499.5	474.5	10	
7	548		499.5	549.5	524.5	21	
8	442		549.5	599.5	574.5	20	
9	663		599.5	649.5	624.5	19	
10	544		649.5	699.5	674.5	11	
11	451		699.5	749.5	724.5	7	
12	508		749.5	799.5	774.5	4	
13	415		799.5	849.5	824.5	0	
14	493						

Copy: **The contents of cell G2 to cells G3:G13**

To construct the cumulative frequency polygon, proceed as in the previous example using values **G2:G13** instead of values **F2:F13**.

	A	B	C	D	E	F	G
1	762		Class Interval		Midpoint	Frequency	Cumulative Frequency
2	433		249.5	299.5	274.5	0	0
3	712		299.5	349.5	324.5	1	1
4	566		349.5	399.5	374.5	2	3
5	618		399.5	449.5	424.5	5	8
6	340		449.5	499.5	474.5	10	18
7	548		499.5	549.5	524.5	21	39
8	442		549.5	599.5	574.5	20	59
9	663		599.5	649.5	624.5	19	78
10	544		649.5	699.5	674.5	11	89
11	451		699.5	749.5	724.5	7	96
12	508		749.5	799.5	774.5	4	100
13	415		799.5	849.5	824.5	0	100

Section 11-2 Measures of Central Tendency

Example 1: Finding the Mean

Enter: **The data in cells A1:A8**

Enter: **The formula =AVERAGE(A1:A8) in cell B10**

	A	B	C
1	3		
2	5		
3	1		
4	8		
5	6		
6	5		
7	4		
8	6		
9			
10	Mean	=AVERAGE(A1:A8)	

Example 2: Finding the Mean for Grouped Data

Enter: **A frequency table for the data as discussed in Section 11-1**

	A	B	C
1	Class Interval	Midpoint x_i	Frequency f_i
2	299.5-349.5	324.5	1
3	349.5-399.5	374.5	2
4	399.5-449.5	424.5	5
5	449.5-499.5	474.5	10
6	499.5-549.5	524.5	21
7	549.5-599.5	574.5	20
8	599.5-649.5	624.5	19
9	649.5-699.5	674.5	11
10	699.5-749.5	724.5	7
11	749.5-799.5	774.5	4

Enter: **The formula =B2*C2 in cell D2**

Copy: **The contents of cell D2 to cells D3:D11**

	A	B	C	D
1	Class Interval	Midpoint x_i	Frequency f_i	Product x_i*f_i
2	299.5-349.5	324.5	1	324.5
3	349.5-399.5	374.5	2	749
4	399.5-449.5	424.5	5	2122.5
5	449.5-499.5	474.5	10	4745
6	499.5-549.5	524.5	21	11014.5
7	549.5-599.5	574.5	20	11490
8	599.5-649.5	624.5	19	11865.5
9	649.5-699.5	674.5	11	7419.5
10	699.5-749.5	724.5	7	5071.5
11	749.5-799.5	774.5	4	3098

Enter: **The formula =SUM(D2:D11)/SUM(C2:C11) in cell A14**

	A	B	C	D
1	Class Interval	Midpoint x_i	Frequency f_i	Product x_i*f_i
2	299.5-349.5	324.5	1	324.5
3	349.5-399.5	374.5	2	749
4	399.5-449.5	424.5	5	2122.5
5	449.5-499.5	474.5	10	4745
6	499.5-549.5	524.5	21	11014.5
7	549.5-599.5	574.5	20	11490
8	599.5-649.5	624.5	19	11865.5
9	649.5-699.5	674.5	11	7419.5
10	699.5-749.5	724.5	7	5071.5
11	749.5-799.5	774.5	4	3098
12				
13	Mean			
14	=SUM(D2:D11)/SUM(C2:C11)			

Example 3: Finding the Median

Enter: **The data in cells A1:A7**

Enter: **The formula =MEDIAN(A1:A7) in cell A10**

	A	B
1	$17,000	
2	$20,000	
3	$28,000	
4	$18,000	
5	$18,000	
6	$120,000	
7	$24,000	
8		
9	Median	
10	=MEDIAN(A1:A7)	

Example 5: Finding Mode, Median, and Mean

Enter: **The data in cells A2:A10**

Enter: **The formula =MODE(A2:A10) in cell B2**

Enter: **The formula =MEDIAN(A2:A10) in cell C2**

Enter: **The formula =AVERAGE(A2:A10) in cell D2**

	A	B	C	D
1	Data	Mode	Median	Mean
2	4	=MODE(A2:A10)		
3	5	MODE(number 1, [number 2], ...)		
4	5			
5	5			
6	6			
7	6			
8	7			
9	8			
10	12			

Note: If a data set has more than one mode (such as the one at the right having mode(s) of 3 and 7), MODE only returns one of these.

Note: If the data set contains no duplicate data points, MODE returns the #N/A error value.

	A	B	C	D
1	Data	Mode	Median	Mean
2	1	3	5	6.090909
3	2			
4	3			
5	3			
6	3			
7	5			
8	6			
9	7			
10	7			
11	7			
12	23			

Section 11-3 Measures of Dispersion

Example 1: Finding the Standard Deviation

Enter: **The data in cells A2:A6**

Enter: **The formula =A2-AVERAGE($A$2-$A$6) in cell B2.**

Copy: **The contents in cell B2 to cells B3:B6.** This subtracts the deviation of each sample measurement from the mean.

	A	B	C	D
1	Data	Dev. Mean	Squared Dev	Sample Variance
2	1	=A2-AVERAGE($A$2:$A$6)		
3	3			
4	5			Sample Std Dev
5	4			
6	3			

Enter: **The formula =B2^2 in cell C2.**

Copy: **The contents in cell C2 to cells C3:C6.** This squares the deviations of each sample measurement from the mean.

	A	B	C	D
1	Data	Dev. Mean	Squared Dev	Sample Variance
2	1	-2.2	=B2^2	
3	3	-0.2		
4	5	1.8		Sample Std Dev
5	4	0.8		
6	3	-0.2		

Enter: **The formula =SUM(C2:C6)/(5-1) in cell D2.** This is the sample variance.

	A	B	C	D
1	Data	Dev. Mean	Squared Dev	Sample Variance
2	1	-2.2	4.84	=SUM(C2:C6)/(5-1)
3	3	-0.2	0.04	
4	5	1.8	3.24	Sample Std Dev
5	4	0.8	0.64	
6	3	-0.2	0.04	

Enter: **The formula =SQRT(D2) in cell D5.** This is the sample standard deviation.

	A	B	C	D
1	Data	Dev. Mean	Squared Dev	Sample Variance
2	1	-2.2	4.84	2.2
3	3	-0.2	0.04	
4	5	1.8	3.24	Sample Std Dev
5	4	0.8	0.64	=SQRT(D2)
6	3	-0.2	0.04	

Alternatively, Excel has a single built-in function **STDEV** for computing the sample standard deviation. The formula **=STDEV(A2:A6)** returns the sample standard deviation of the data set as well. To compute the population standard deviation, use **STDEVP**.

Computing the sample variance and the population variance can be done using the **VAR** and **VARP** functions.

Example 2: Finding the Standard Deviation for Grouped Data

Enter: **The data in cells A2:A6 and their corresponding frequencies in cells B2:B6**

Enter: **The formula =A2-AVERAGE($A$2-$A$6) in cell C2.**

Copy: **The contents in cell C2 to cells C3:C6.**

	A	B	C	D	E
1	Data	Frequency	Dev Mean	Squared Dev	
2	8	1	=A2-AVERAGE($A$2:$A$6)		
3	9	2			
4	10	4			
5	11	2			
6	12	1			

Enter: **The formula =B2*C2^2 in cell D2.**

Copy: **The contents in cell D2 to cells D3:D6.** The sum of the contents of column **D** is the sum of the squared deviations of the data set.

	A	B	C	D
1	Data	Frequency	Dev Mean	Squared Dev
2	8	1	-2	=B2*C2^2
3	9	2	-1	
4	10	4	0	
5	11	2	1	
6	12	1	2	

Enter: **The formula =SUM(D2:D6)/(SUM(B2:B6)-1) in cell E2.** This is the sample variance.

	A	B	C	D	E	F
1	Data	Frequency	Dev Mean	Squared Dev	Sample Variance	
2	8	1	-2	4	=SUM(D2:D6)/(SUM(B2:B6)-1)	
3	9	2	-1	2		
4	10	4	0	0	Sample Std Dev	
5	11	2	1	2		
6	12	1	2	4		

Enter: **The formula =SQRT(E2) in cell E5.** This is the sample standard deviation.

	A	B	C	D	E
1	Data	Frequency	Dev Mean	Squared Dev	Sample Variance
2	8	1	-2	4	1.333333333
3	9	2	-1	2	
4	10	4	0	0	Sample Std Dev
5	11	2	1	2	=SQRT(E2)
6	12	1	2	4	

Section 11-4 Bernoulli Trials and Binomial Distributions

Example 3: Probability of x Successes in n Bernoulli Trials

Enter: **The values of n (number of repeated Bernoulli trials), p (probability of success) in cells A2 and B2.**

Enter: **The formula =1-B2 in cell C2.**

	A	B	C
1	n	p	q
2	5	0.166666667	=1-B2

Enter: **The value 0 in cell A5**

Enter: **The formula =A5+1 in cell A6**

Copy: **The contents of cell A6 to cells A7:A10**

Enter: **The formula =COMBIN($A$2,A5)*$B$2^A5*$C$2^($A$2-A5) in cell B5.**

	A	B	C	D	E
1	n	p	q		
2	5	0.166666667	0.833333		
3					
4	x	Pr(x successes)			
5	0	=COMBIN($A$2,A5)*$B$2^A5*$C$2^($A$2-A5)			
6	1				
7	2				
8	3				
9	4				
10	5				

Copy: **The contents of cell B5 to cells B6:B10**

	A	B	C
1	n	p	q
2	5	0.166666667	0.833333
3			
4	x	Pr(x successes)	
5	0	0.401877572	
6	1	0.401877572	
7	2	0.160751029	
8	3	0.032150206	
9	4	0.003215021	
10	5	0.000128601	

The probability of exactly two successes can be found in cell **B7**. To compute the probability of at least two successes, use the formula **=SUM(B7:B10)** or the formula **=1-SUM(B5:B6)**.

Example 7: Patient Recovery

Enter: **The values of *n* (number of repeated Bernoulli trials), *p* (probability of success) in cells A2 and B2.**

Enter: **The formula =1-B2 in cell C2.**

Enter: **The value 0 in cell A5**

Enter: **The formula =A5+1 in cell A6**

Copy: **The contents of cell A6 to cells A7:A10**

Enter: **The formula =COMBIN($A$2,A5)*$B$2^A5*$C$2^($A$2-A5) in cell B5.**

	A	B	C	D	E
1	n	p	q		
2	8	0.5	0.5		
3					
4	x	Pr(x successes)			
5	0	=COMBIN($A$2,A5)*$B$2^A5*$C$2^($A$2-A5)			
6	1				
7	2				
8	3				
9	4				
10	5				
11	6				
12	7				
13	8				

Copy: **The contents of cell B5 to cells B6:B13**

	A	B	C
1	n	p	q
2	8	0.5	0.5
3			
4	x	Pr(x successes)	
5	0	0.00390625	
6	1	0.03125	
7	2	0.109375	
8	3	0.21875	
9	4	0.2734375	
10	5	0.21875	
11	6	0.109375	
12	7	0.03125	
13	8	0.00390625	

Choose: **Chart Wizard > Column >**

Select: **Chart sub-type: Clustered Column.**

Click: **Next**

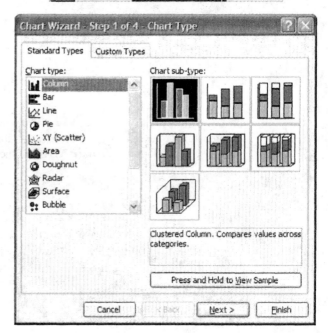

Select: **Data Range Tab**

Enter: Data range: **B5:B13**

Select: Series in: **Columns**

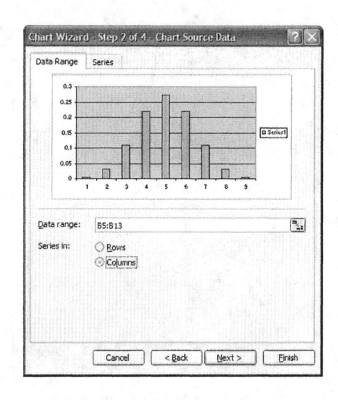

Select: **Series Tab**

Select: Category (X) axis labels: by clicking on source data square at right of entry box.

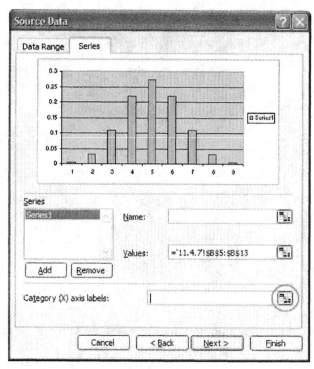

Select: **Cells A5:A13 using the mouse.**

Click: **Circled box on right of Source Data box**

Click: **Next**

	A	B	C	D	E	F
1	n	p	q			
2	8	0.5	0.5			
3						
4	x	Pr(x successes)				
5	0	0.00390625				
6	1	0.03125				
7	2	0.109375				
8	3	0.21875				
9	4	0.2734375				
10	5	0.21875				
11	6	0.109375				
12	7	0.03125				
13	8	0.00390625				
14						
15						
16	='11.4.7!$A$5:$A$13					
17						

Chart Wizard - Step 2 of 4 - Chart Source Data - Catego...

Select: **Data Labels Tab**

Select: Label Contains: **Value**

Select: **Legend Tab**

De-select: **Show Legend**

Select: **Titles Tab**

Enter: Category (X) axis: **Number of successes, x**

Enter: Value (Y) axis: **P(x)**

Click: **Finish**

Right-click: on any data label on the graph

Select: **Format Data Labels…**

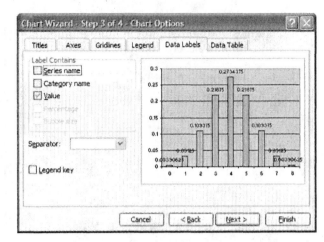

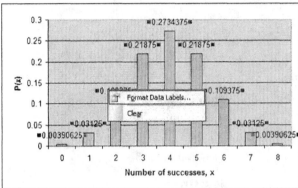

Select: **Number Tab**

Select: Category: **Number**

Enter: Decimal places: **3**

Click: **OK**

Right-click: on any bar on the graph

Select: **Format Data Series...**

Select: **Options Tab**

Enter: Gap width: **0**

Click: **OK**

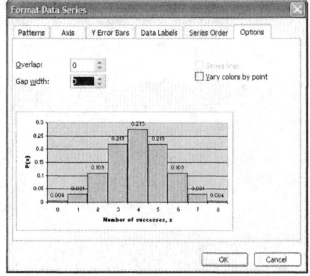

To compute the mean and standard deviation of this distribution, use the formulas **=A2*B2** and **=SQRT(A2*B2*C2)** respectively.

Section 11-5 Normal Distribution

Example 1: Finding Probabilities for a Normal Distribution

Enter: **The value 500 (the mean) in cell A2**

Enter: **The value 100 (the standard deviation) in cell B2**

Enter: **The formula =NORMDIST(670,A2,B2,TRUE)-NORMDIST(500,A2,B2,TRUE) in cell C2**

	A	B	C	D	E	F	G	H
1	Mean	Std. Dev.	Prob.					
2	500	100	=NORMDIST(670,A2,B2,TRUE)-NORMDIST(500,A2,B2,TRUE)					

Note: The option TRUE above is used to return the cumulative distribution function. So, the formula is computing the probability of being less than 670 and subtracting the probability of being less than 500.

Example 2: Finding Probabilities for a Normal Distribution

Enter: **The value 500 (the mean) in cell A2**

Enter: **The value 100 (the standard deviation) in cell B2**

Enter: **The formula =NORMDIST(500,A2,B2,TRUE)-NORMDIST(380,A2,B2,TRUE) in cell C2**

	A	B	C	D	E	F	G	H
1	Mean	Std. Dev.	Prob.					
2	500	100	=NORMDIST(500,A2,B2,TRUE)-NORMDIST(380,A2,B2,TRUE)					

Example 3: Market Research

Enter: **The values of *n* (number of repeated Bernoulli trials), *p* (probability of success) in cells A2 and B2.**

Enter: **The formula =1-B2 in cell C2.**

Enter: **The formula =A2*B2 in cell A5**

Enter: **The formula =SQRT(A2*B2*C2) in cell B5**

	A	B	C
1	n	p	q
2	20	0.4	0.6
3			
4	Mean	Std. Dev.	Prob.
5	8	=SQRT(A2*B2*C2)	

Enter: **The formula =NORMDIST(12.5,A5,B5,TRUE)-NORMDIST(5.5,A5,B5,TRUE) in cell C5**

	A	B	C	D	E	F	G	H
1	n	p	q					
2	20	0.4	0.6					
3								
4	Mean	Std. Dev.	Prob.					
5	8	2.19089	=NORMDIST(12.5,A5,B5,TRUE)-NORMDIST(5.5,A5,B5,TRUE)					

To approximate the probability that the sample contains fewer than 4 users of the credit card,

Enter: **The formula =NORMDIST(3.5,A5,B5,TRUE) in cell C5**

	A	B	C
1	n	p	q
2	20	0.4	0.6
3			
4	Mean	Std. Dev.	Prob.
5	8	2.19089	0.01999